AF704535

MEMS Cost Analysis

MEMS Cost Analysis

From Laboratory to Industry

Ron Lawes

Pan Stanford Publishing

Published by

Pan Stanford Publishing Pte. Ltd.
Penthouse Level, Suntec Tower 3
8 Temasek Boulevard
Singapore 038988

Email: editorial@panstanford.com
Web: www.panstanford.com

British Library Cataloguing-in-Publication Data
A catalogue record for this book is available from the British Library.

MEMS Cost Analysis: From Laboratory to Industry

Copyright © 2014 Pan Stanford Publishing Pte. Ltd.

All rights reserved. This book, or parts thereof, may not be reproduced in any form or by any means, electronic or mechanical, including photocopying, recording or any information storage and retrieval system now known or to be invented, without written permission from the publisher.

For photocopying of material in this volume, please pay a copying fee through the Copyright Clearance Center, Inc., 222 Rosewood Drive, Danvers, MA 01923, USA. In this case permission to photocopy is not required from the publisher.

ISBN 978-981-4411-06-6 (Hardcover)
ISBN 978-981-4411-07-3 (eBook)

Printed in the USA

To my wife, Jean, who has listened for 54 years (and counting) to my obsession with technology. It would not have been possible to write this book if she had not fed me, watered me and corrected some of the text, but mainly had she not patiently waited several years for me to address numerous household jobs.

Contents

Preface

Manufacture of microelectromechanical systems (MEMS) is a major industry providing products to the consumer, automotive, medical and defence sectors. The products are small, often occupying an area less than 1 mm^2, have a high degree of functionality and are relatively cheap to manufacture (less than $1 per device).

Only 20 years ago, MEMS was mainly an academic activity, with many of the fabrication techniques and devices themselves in research or, at best, development. There seemed to be as many different methods of manufacture as there were devices. Nowadays there has been much standardisation in both fabrication, with silicon MEMS becoming mainstream, and packaging, where much has been adapted from the semiconductor industry. The cost of manufacture, including fabrication, testing and packaging, has become a major issue.

This volume seeks to show how cost analysis can be applied to MEMS, taking into account the wide range of processes and equipment, the major differences with the established semiconductor industry and the presence of both large-scale, product-orientated manufacturers and small- and medium-scale foundries. It examines the processes and equipment sufficiently for the reader to appreciate how costs arise. Appropriate publications are referenced so that technical details may be examined, outside the confines of cost analysis. Representative costs for equipment, processes and some products are examined in sufficient depth to show how financial models can be introduced to estimate the cost and price for a MEMS product.

Chapter 1 provides the historical background to the growth of the MEMS industry from its origins in the more mature semiconductor industry. Some of the key milestones over the last 50 years are noted. Chapter 2 introduces basic fabrication processes that are fundamental to MEMS, notably material deposition, patterning and etching.

Chapter 3 introduces the equipment typical of surface micro-engineering of a few microns' depth, while Chapter 4 looks at high-aspect-ratio techniques where the microengineering can be several hundred microns in depth.

Chapter 5 examines the basis for testing MEMS devices, noting that this is a technology requiring equipment development. Chapter 6 discusses techniques to package the MEMS and signal-processing requirements, many of which come from the semiconductor industry.

MEMS manufacture must be carried out in ultra-clean rooms, where the particulate count in the air must be low and the size of particles sub-micron. The clean rooms, the manufacturing equipment and the plant to supply consumables must be installed in a building shell. This requires considerable investment that ultimately must be added to any cost estimate for a product. These issues are covered in Chapter 7.

Chapter 8 shows the structure of one version of a cost spreadsheet (MEMSCOST, written by the author) that brings together the technical details of MEMS fabrication, testing and packaging, with financial parameters, sufficient to cost any MEMS product.

Some examples of commercially successful MEMS products are examined in detail. Chapter 9 examines the cost of fabricating, testing and packaging three-axis accelerometers over a period of years, as predicted by MEMSCOST and compared with publically available market data. Chapter 10 repeats the process for the examination of MEMS microphones.

Commercial foundries have become an important source of manufacturing technology for the industry. Chapter 11 investigates the operation of foundries, contrasting multi-use foundries with single-product integrated device manufacturers (IDMs), some of which offer a job-shop capability and others (e.g., Sandia, MEMSCAP) a fixed process technology for a fixed price.

The purpose of MEMSCOST is to understand and calculate the cost of producing MEMS devices, to enable better investment and to make pricing decisions. Chapter 12 looks at the overall financial analysis.

The methodology outlined has its origins as early as the 1990s, when the author was the founding director, Central Microstructure Facility, Rutherford and Appleton Laboratories, UK, from 1977 to 2003. In recent years this methodology has been further developed by the author to include a wider range of applications and more developed accounting procedures.

Professor Ron Lawes
Fellow, Royal Academy of Engineering
June 2012

Acknowledgements

I am indebted to many colleagues from the industry and academia for discussing MEMS technology and costs with me over the years and who have helped me to arrive at an understanding of industrial costs.

My colleague Professor Richard Syms at Imperial College London helped with the various design and fabrication aspects of accelerometers and microphones, as well as providing me with many photographs. Professor Andrew Holmes and Professor Malcolm Gower discussed many aspects of lasers and related technology.

Professors Anthony Walton and Tom Stevenson from the Scottish Microfabrication Centre at the University of Edinburgh and Professor Michael Kraft and Dr Iain Anteney from the Southampton Nanofabrication Centre at the University of Southampton provided excellent photographs of their clean-room facilities and equipment. Dr Mike Ward of Birmingham University provided an excellent photograph of an etched microstructure.

Professor Rob Santilli (AML) introduced me to wafer-bonding technology and showed me the cost structure of a typical small- and medium-sized company. Clive Bond at Loadpoint pointed out some of the finer points of wafer dicing and provided photographs of their equipment.

Dr Erol Harvey and Michael Wilkinson at MiniFAB, Australia, brought me into an industrial environment from a career in a government research laboratory and thus provided the industrial background that went into much of this book. MiniFAB also supplied data and pictures. Dr Matteo Altissimo from Monash University, Australia, supplied me with a picture of their electron beam machine.

Professor Volker Saile at Karlsruhe gave me the opportunity to study the potential for LIGA technology and provided a photograph of the synchrotron experimental area. The Diamond Light Source, UK, supplied an aerial view of the Diamond Synchrotron.

I am indebted to 'Sinjin' St John Dixon-Warren from Chipworks for the magnificent pictures of accelerometers and microphones shown in Chapters 9 and 10, respectively.

I also wish to acknowledge the roll of my former employer, the Science and Technology Facilities Council at the Rutherford Appleton Laboratory, for encouraging me to spend 42 years of research enjoying my passionate interest in technology.

A key analysis tool has been my MEMSCOST spreadsheet, which was developed to include details of MEMS processes and equipment and has enabled cost and other financial analysis. My daughter, Catherine, provided guidance on aspects of financial accounting and proofread every chapter.

However, any errors of fact or computation are mine and mine alone.

Chapter 1

The World of MEMS

1.1 Introduction

The ubiquitous semiconductor device, often referred to as a silicon chip or integrated circuit (IC), is manufactured from one of two technologies, namely, the complementary-metal-oxide semiconductor (CMOS) and the bipolar-junction transistor. CMOS is used to manufacture digital logic, and the bipolar-junction transistor is used to manufacture analogue components, such as amplifiers for signal processing. A hybrid technology has been developed from CMOS and the bipolar transistor and is known as BiCMOS.

Initially, large-scale production of CMOS/BiCMOS was limited to standardised circuits such as the central processing unit (CPU) and memory components needed for the personal computer. An important development over the last 30 years has been the application-specific integrated circuit (ASIC), where, as the name suggests, an IC design is customised for a particular application. The demand for ASICs has led to the establishment of IC foundries, discussed in Chapter 11.

Over the last 20 years a new industry has grown to mimic the IC industry and is known as microelectromechanical systems (MEMS) or microsystems technology (MST). ICs contain only electronic

MEMS Cost Analysis: From Laboratory to Industry
Ron Lawes
Copyright © 2014 Pan Stanford Publishing Pte. Ltd.
ISBN 978-981-4411-06-6 (Hardcover), 978-981-4411-07-3 (eBook)
www.panstanford.com

components with features on a sub-100 nm scale, within a chip size a few hundred square millimeters. In contrast MEMS contain mechanical components with features on a micron scale and within a die size of a few square millimetres (Bryyzek, 2005; 2006). A typical MEMS device, for example, an accelerometer or a microphone will have a BiCMOS signal-processing ASIC connected to a MEMS mechanical sensor.

The key developments in MEMS are shown in Table 1.1 and consist of:

(1) early predictions of the potential of MEMS by luminaries such as Nobel Laureate Richard Feynman (Feynman, 1959; Feynman, 1993) and the identification of silicon as having key mechanical properties by Kurt Peterson (Peterson, 1982).
(2) the extension of known silicon-processing techniques from the semiconductor industry, such as wet silicon etching, metal deposition and the development of new application-specific processes. Examples include surface and bulk micromachining, deep reactive-ion etching (DRIE), wafer bonding and the use of sacrificial layers. Processes for non-silicon materials, for example ultraviolet (UV) laser technology for micromachining polymers, have been developed.
(3) the development of MEMS microcomponents needed for system integration, for example cantilevers, comb actuators, pressure sensors, digital mirrors and spatial light modulators.
(4) the development of MEMS systems on a single chip, such as miniature accelerometers, gyros, microphones, oscillators, biosystems and optical switching networks.
(5) the development of infrastructure, such as design and simulation software (e.g., Coventor) and MEMS foundries (e.g., IMT, Silex, Tronics; see Chapter 11) to support the growing industry.

MEMS devices are increasingly used in many industries, including automotive, consumer, communications, defence and aerospace. Major MEMS products include inkjet printer heads, accelerometers (Eloy, 2011), gyros, pressure sensors, microphones, radio frequency (RF) switches, microfluidics and microscanning mirrors. The rate of development of new versions of these products has been outstanding. For example, the MEMS accelerometer has developed

Table 1.1 Principle MEMS events 1959–2010

	Item	Source
1959	'There's plenty of room at the bottom'	R Feynman
1959	silicon strain gauge	Kulite
1960	Surface micromachining	H Nathanson
1969	Anisotropic wet etch (KOH)	HA Waggener
1979	Thermal inkjet printer	HP, IBM
1982	Silicon as a mechanical material	K Peterson
1982	LIGA	Karlsruhe
1982	Laser micromachining	K Jain, Lawes, Goodall
1982	Blood pressure sensor	Novasensor, Motorola, EG&G
1983	'Infinitesimal machinery'	R Feynman
1983	Integrated pressure sensor	Honeywell
1985	Auto engine pressure sensor	TRW
1985	Airbag sensor	SensorNor, ADI
1986	Silicon wafer bonding	M Shimbo
1987	1st generation accelerometer	TRW, Lucas NovaSensor
1987	Digital mirror (R&D)	Hombeck TI
1988	Electrostatic drive motors	Berkeley, Fan, Tai, R Muller
1989	Lateral comb drive actuators	Berkeley
1989	Si anisotropic wet etch (TMAH)	
1992/3	PolyMUMPS	MCNC, Cronos
1992	Bulk micromachining (SCREAM)	Cornell
1992	1st MEMS hinge	K Pister, Judy, Burgett, Fearing
1992	Grating light modulator	Solgaard, Sandejas, Bloom
1993	Integrated accelerometer	Analog Devices.
1994	DRIE	Bosch process patented
1996	Digital mirrors	TI's 1st commercial product
1998	SUMMIT process	Sandia
1999	Optical network switches	Lucent
1999	Corner cube reflector	Hsu
2000	Silicon microphones	
2001	MEMS oscillators	SciTime
2003/5	Dual-axis gyroscope	S Nasri (InvenSense)
2006	Accelerometers	STMicroelectonics enter the market
2007	Mass storage terabits	IBM Millipede
2008	Single-axis gyro	STMicroelectronics
2009	2-axis gyro	STMicroelectronics
2010	3-axis gyro	STMicromicroelectronics

Abbreviations: R&D, research and development; LIGA, Lithographie, Galvanoformung and Abformung; TMAH, tetramethylammonium hydroxide; SCREAM, single-crystal reactive etching and metallization; SUMMIT, Sandia ultra-planar, multi-level MEMS technology; KOH, potassium hydroxide.

from a single-axis version in 1997 to a two-axis version in 2000 to the present three-axis version, all in a similar package.

Inkjet printer cartridges have been a major MEMS area (Singh et al., 2009), where the intricate structure of ink channels is formed by microengineering, that is, a branch of microfluidics. The use of MEMS technology is exemplified by manufacture of the thermal inkjet cartridge, where two types of construction prevail. HP and Canon fabricate a cartridge with an integral ink reservoir, microfluidic channels and a resistive heater. A bubble of ink is formed by heating and ejection from the microchannel on to the paper. Epson has pioneered a different technique, whereby a piezoelectric actuator is deformed by a current pulse, which then forces a bubble of ink to be ejected on to the paper.

Both accelerometers and, more recently, gyros have made a major contribution to automobile safety, the accelerometer as the crash sensor to trigger airbag deployment and the gyro for roll sensing (Ernst, 2005). In recent years, they have developed from single-axis devices per package to today's three-axis versions. Other applications of accelerometers and gyros include computers, tablets and mobile telephones. These consumer products have also increased the demand for high-performance, low-cost microphones. Silicon-based MEMS microphones offer such technical and financial benefits with the advantage of being compatible with a CMOS signal processor. In 2006, the digital microphone was introduced, and since then it has been much reduced in size and cost. An example of the use of these three devices is the IPhone 4, which uses a three-axis MEMS accelerometer and a gyro (pitch, roll and yaw) and three MEMS microphones (Chipworks, 2010).

One of the earliest demonstrations of microfluidic components was on silicon, as microfluidic devices require more area (real estate) than other MEMS. Applications for both research and products (MiniFAB, 2011) use predominately non-silicon technologies. Fabrications on glass by lithography, or on polymers by laser ablation, micromilling, embossing or injection molding, are becoming the main cost-effective processes. Microfluidic devices incorporate both passive and active components that are most cost-effective when using the minimum number of processing steps. Active components, including pumps, valves and mixing actuators, require external power sources to drive them, so they are more complex to integrate. Passive components include mixers, separators and filters and in optimal designs are fabricated by the same process that creates the channels and flow reservoirs. Applications include medical equipment such as deoxyribonucleic acid (DNA) microarray chips, lab-on-a-chip devices for diagnostics, point-of-care diagnostics, microreactors and fundamental biodiscovery applications. Microfluidics, aside from the inkjet cartridge described above, is expected to be one of the most significant market areas of MEMS by 2015.

Silicon digital-controlled micromirrors (DLMs) have been applied to several fields of display technology, notably in large projection TV systems (Texas Instruments). In recent years different MEMS-based display technologies have been developed for 'hand-held' projectors or pico-projectors. Pixtronix has developed a technology based on digital microshutters (DMSs), whereby red, green and blue light is modulated by MEMS shutters (Pixtronix, 2011). Time-multiplexed optical shutter (TMOS) technology also uses red, blue and green light, but modulation is achieved by capacitive plates, which, when brought together, enable light to be transmitted (unipixel). The PicoP system uses a MEMS scanner to write an image, pixel by pixel, for each line of the image. The scanner contains the necessary laser light sources, optics and electronics (Microvision, 2011).

MEMS pressure sensors are based upon a thin membrane, which deflects under a change of pressure and then causes a change in capacitance between two plates. Pressure sensors find many applications, particularly in the automobile industry, and will continue to grow over the next few years into one of the major application areas for MEMS (Dixon, 2011). The automotive industry uses MEMS pressure and temperature sensors, particularly for the harsh environment of the engine compartment. Another major area of application is in the medical market, where miniature pressure sensors are part of an array of sensors on catheters (Rebello, 2004).

In terms of market share RF MEMS is a small market, which is expected to grow significantly over the next two to three years from $25 million to $225 million by 2014 (Bouchaud, 2010). RF MEMS resonators are poised to replace the quartz oscillator in timing and

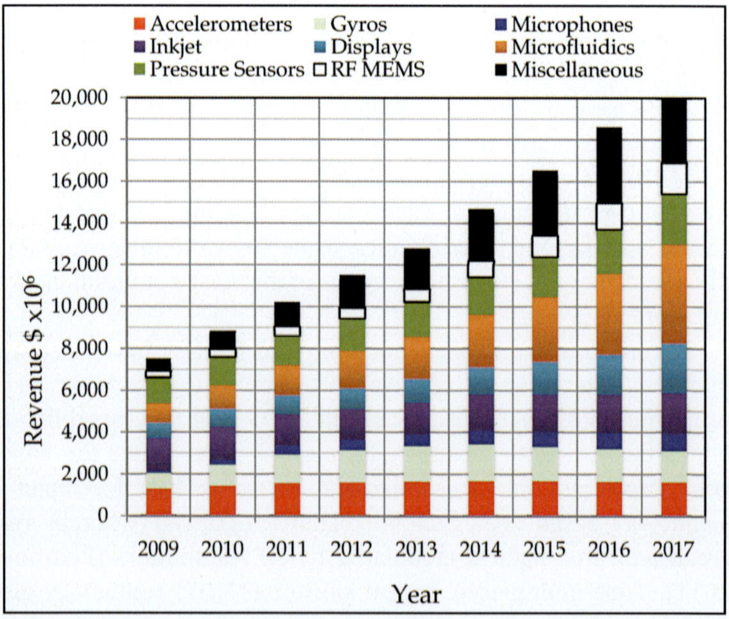

Figure 1.1 Major application areas for MEMS technology (adapted from Mounier, 2012).

clock applications (SiTime, 2009). Such devices are based on silicon and will enable low cost, small size and integration with CMOS.

RF MEMS switches and RF MEMS varactors (variable capacitance) also offer improved performance, low cost, small size and integration with CMOS for wireless front-ends. Application areas will include communications (e.g., mobile telephones), medicine, aerospace and the automotive industry (Varadon et al., 2003; Lucyzyn, 2004; Bouchard, 2010).

Figure 1.1 shows the magnitudes of these market areas in the immediate past with estimates for the future up to 2017 (Mounier, 2012). The total revenue for the MEMS market in 2011 was estimated at approximately $10 billion (compared with the semiconductor industry estimated at over $300 billion), of which some 80% was confined to just seven application areas (Eloy, 2011). Note that other important applications, such as RF MEMS, microbolometers, micro-opto-electromechanical systems

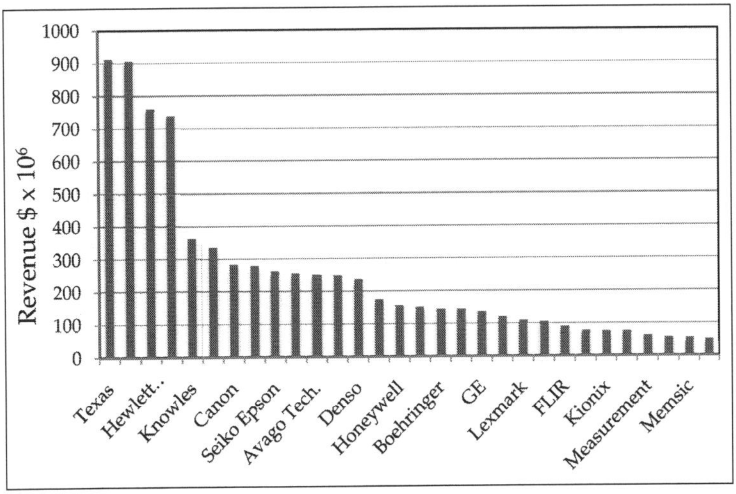

Figure 1.2 Revenue for MEMS companies, 2011 (estimated from Yole, 2012).

(MOEMS) and digital compasses, are included in the 'miscellaneous' designation.

Figure 1.2 shows the revenue from the first 30 highest-ranking MEMS companies (Yole, 2011). The average revenue per company was estimated at $250 million. To demonstrate the methodology of cost analysis for MEMS devices and to enable strong correlation with market data, Chapters 9 and 10 will concentrate on the cost analysis of accelerometers and microphones, respectively.

Significant MEMS production was first started on 100 mm diameter wafers and then migrated to 150 mm wafers as production issues, such as cost, became more important. Currently, the majority of the world's MEMS foundries manufacture on 150 mm/200 mm diameter wafers, with the large manufacturers using 200 mm and looking to 300 mm fabrication to reduce costs even further. At each stage of the migration to larger wafers, cost is a driving factor, and cost analysis is a vital tool needed to assess any financial advantages.

MEMS and ICs have many similarities in fabrication but have different principles of operation. The principles of an IC are based on semiconductor physics (classical and quantum), whereas a MEMS device is based predominately on mechanical principles and may also employ optical, acoustical, thermal, piezoelectric and magnetic principles. When the MEMS device requires signal processing to function, it is connected in some manner to a specialist ASIC.

Some MEMS can be manufactured from the same materials as the IC, predominately silicon and its variants. However, the development of MEMS has prospered from the use of a much wider range of materials, such as metals and polymers. The IC is a planar device, whereas MEMS is three-dimensional. The IC consists of multiple sub-micron layers, whereas a MEMS device can have significant depth (5–500 μm) required for its operation, for example, to gain signal strength.

In analysing the costs of a MEMS device, the manufacturing process will be divided into three parts, namely, *fabrication*, *testing* and *packaging*. There is a tendency for the MEMS industry to develop a unique process for each product, which makes it difficult to achieve standardisation and low cost. In contrast, products based on silicon and using predominately surface microengineering will have significant advantages in both process and equipment standardisation and potentially enable larger volumes and second-source manufacture.

Much of the equipment and techniques needed to fabricate a MEMS device, such as lithography and the deposition and etching of thin films of material (Chapter 2), come directly from the semiconductor industry. A selection of equipment common to both MEMS and IC fabrication is described in Chapter 3. The need for three-dimensional fabrication, with near-vertical walls, has led to the development of special equipment and techniques for MEMS, such as DRIE, excimer laser and wafer bonding. Some MEMS devices are sensitive to traces of liquids after processing, where any resultant capillary action can prevent mechanical movement. The equipment and processes are described in Chapter 4 under critical point damping.

It is in the areas of testing and packaging that some notable differences occur, and it is difficult to generalise about the testing and packaging of MEMS. For example, semiconductor device tests predominately check the electronic and functional logic performance of a device, whereas a MEMS device may need to be checked in its response to acceleration, pressure, temperature, etc. Testing is discussed in Chapter 5.

This leads to a problem with packaging. In some cases, for example, accelerometers, the device must be hermetically sealed from the environment, and in the case where a CMOS/BiCMOS die is included in the package, the MEMS die must also be protected before further packaging processes can be carried out. The solution is to cap the MEMS device (Chapter 6), and several methods have been developed. Packages, packaging equipment and processes, developed for the semiconductor industry, can be used for these MEMS devices. For other MEMS devices, connection to the outside environment is the essence of the device and hence novel or modified packages have to be developed.

The buildings and the plant that are needed to house an IC plant now cost billions of dollars, mainly for the construction and operation of ultra-clean rooms, ultra-pure de-ionised water and chemicals. Such costs can be amortised into the production costs and are discussed in Chapter 7.

As MEMS technology has matured, along with the development of specialist foundries to manufacture prototypes, the cost of manufacturing volume products has become critical. For example, an error of only $+/-1\%$ in the unit cost for a typical MEMS product, manufactured in large volumes, could result in a \$1 million movement in profit or loss. For every company, accurate cost analysis is needed to ensure that design and manufacture can meet the market price and to provide management with information on the cost-effectiveness of various strategies, such as pre- and post-packaging testing.

IC cost analysis has been available commercially for many years and has been continually improved in scope and accuracy. Extension of cost analysis to MEMS has been slow, as the processes and equipment have been much more varied than is the case with ICs. However, in recent years there has been some consolidation in MEMS processes and equipment, so useful cost models can be developed. An example of a typical methodology is the spreadsheet MEMSCOST designed by the author. This is used to cost MEMS manufacture and to demonstrate the variation in costs arising from different design assumptions, such as the capital cost and equipment needed to meet a given annual output. MEMSCOST calculates fixed and variable costs and enables the calculation of revenue and profit. MEMSCOST is written in Microsoft Excel 2008, currently occupies 5 MB and is discussed in Chapter 8.

The technical details of the design and fabrication of MEMS devices can be found in the various references and bibliography associated with each chapter. In this volume, two examples of commercial MEMS devices will be analysed sufficiently to illustrate the details of cost analysis. The first device will be the three-axis accelerometer, a device that has been developed over several decades and for which much technical and financial data is available. In particular, the cost of packaging will be compared with fabrication and testing (Chapter 9).

The second device will be the MEMS microphone, which also has technical and financial data that can be compared with commercially available data (Chapter 10).

In the 1980s, the cost of manufacturing the next generation of ICs became prohibitively high and some companies looked for alternatives—hence the development of the CMOS foundry where standard CMOS processes were offered to different customers. Today, companies such as Taiwan Semiconductor Manufacturing Company (TMSC, Taiwan, 1987), United Microelectronics Corporation (UMC, Taiwan, 1980) and Global Foundries (Dresden, 2009) dominate the production of ICs outside the in-house manufacturers such as Intel.

A similar phenomenon has occurred with MEMS (see Chapter 11) as such devices become essential for advanced systems, for example, iPods and telephones. Industrial manufacturers looked for access to a cheap way of manufacturing devices in volume, yet universities and research institutes only require a few die for specific applications. A problem for potential MEMS foundries is that there is relatively little process standardisation unlike the case with CMOS, so the growth of MEMS foundries has been more varied and smaller. Revenue for leading foundries such as Silex and IMT is \$30–\$40 million, compared with \$800–\$900 million for an in-house MEMS manufacturer.

A significant development has been the design of key MEMS devices based entirely on silicon as a basic material and CMOS-like processes and equipment. This has had two major consequences. Firstly, several MEMS foundries have been converted from obsolete CMOS factories as the minimum dimensions are less demanding (e.g., 1 μm compared with 0.1 μm), and secondly the large CMOS foundries can readily enter the silicon MEMS market (e.g., Asia Pacific, TMSC), with the resultant implications for cost.

References

Bouchaud, J (2010). RF MEMS switches and varactors set to deliver on their promise. *MEMS Special Report, IHS iSuppli.*

Bryzek, J (2005). Principles of MEMS. *Handbook of Measuring System Design.* John Wiley & Sons.

Bryzek, J, et al. (2006). Marvelous MEMS advanced IC sensors and microstructures for high volume applications. *IEE Circuits Devices Mag.*, **22**(2), 8–28.

Chipworks (2010). Teardown of the Apple iPhone4. www.chipworks.com/.../silicon-teardown-of-the apple-iphone-4-smart-phone

Dixon, R (2011). Pressure sensors to become top MEMS device by 2014. *IHS Suppli Press Release*, October **27**, 2011.

Ernst, P (2005). MEMS @ Bosch: automotive applications and beyond. www.MSTBW.de/... / bosch_mems_12_micromachine_symposium_ernst.pdf

Eloy, JC (2011). Inertial combo sensors for consumer and automotive. *Yole Development.*

Eloy, JC (2012). STMicroelectronics challenges Texas Instruments for top spot. *Yole Development.*

Feynman, R (1959). There's plenty of room at the bottom. *Cal. Eng. Sci.*, **23**(5), 22–36.

Feynman, R (1993). Infinitesimal machinery. *J. Micromech Syst.*, **2**(1), 4–14.

Lucyzyn, S (2004). Advanced RF MEMS. *Cambridge RF and Microwave Engineering Series.* Cambridge Press.

Microvision (2011). www.microvision.com

MiniFAB (2011). Technology and processes. www.minifab.com.au

Mounier, E (2012). MEMS markets and applications—dMEMS 2012. dmems.univ-fcomte.fr/presentation/mounier.pdf.

Peterson, K (1982). Silicon as a mechanical material. *Proc. IEEE* **70**(5), 420–457.

Pixtronix (2011). www.pixtronix.com

Rebello, KJ (2004). Applications of MEMS in surgery. *Proc. IEEE*, **92**(1).

Singh, M, et al. (2009). Inkjet printing—process and its applications. *Adv. Mater.*, **22**(6), Wiley Online Library.

SiTime (2009). MEMS replacing quartz oscillators. *SiTime Corporation SiT-AN10010 Rev.1.1.*

Unipixel (2011). www.unipixel.com

Varadon, VK, et al. (2003). *RF MEMS and Their Applications.* John Wiley & Sons.

Chapter 2

Basic Fabrication Processes

2.1 Introduction

Over the last 40 years or so, the semiconductor industry has developed manufacturing processes and equipment suitable for the batch processing of silicon. These semiconductor techniques have produced devices of increasing miniaturisation and increasing performance, all at a greatly reduced cost (Levinson, 2011).

These developments have been driven by

(i) multiple devices (from hundreds to thousands), which are fabricated simultaneously in the form of processed dies on a single substrate known as a wafer. Batch processing leads to a much reduced unit cost.

(ii) the use of silicon as the wafer material and the subsequent development of compatible materials for fabrication, such as silicon oxide, silicon dioxide (SiO_2) and polysilicon to form transistors and other microelectronic components.

(iii) a process known as lithography, whereby a three-dimensional microstructure is constructed from multiple layers of two-dimensional patterns. Resolution and alignment of the features, from one layer to another, become more demanding as feature sizes are reduced.

MEMS Cost Analysis: From Laboratory to Industry
Ron Lawes
Copyright © 2014 Pan Stanford Publishing Pte. Ltd.
ISBN 978-981-4411-06-6 (Hardcover), 978-981-4411-07-3 (eBook)
www.panstanford.com

(iv) the selective deposition and etching of material through the windows left by the lithography process.
(v) the separation of the individual chips, now often referred to as 'die', from the wafer.
(vi) the assembly of the die into a durable ceramic or plastic container, a process known as packaging.
(vii) the automatic testing of the device before and/or after packaging.

Microelectromechanical systems (MEMS) fabrication technology uses some of the well-established thin-film processes (a few microns in thickness) and equipment as used in the semiconductor industry (Wolf & Tauber, 1986; 1990). Often referred to as 'surface microengineering' many of the semiconductor industry's thin-film processes are used in MEMS fabrication and are outlined in Chapter 3. However, MEMS designs often require microstructures of hundreds of microns in depth, rather than a few microns, in order to achieve the desired performance, for example, enhanced sensitivity for a sensor. These techniques are referred to as 'high aspect ratio', or HAR, microstructures and have required new processes and equipment to be developed (Lawes, 1996). The most important of these are outlined in Chapter 4.

Silicon and its variants are the most common materials used in the manufacture of MEMS. Typically, a MEMS design in silicon will include thin films of polysilicon or silicon nitride for mechanical structures and SiO_2 as a 'sacrificial layer'. These layers will often be deposited uniformly across the whole wafer ready for patterning. The concept of a sacrificial layer is fundamental to the microfabrication of mechanical structures as it initially forms a firm foundation on which a mechanical structure will be fabricated. The sacrificial layer is removed at a later stage in the fabrication sequence so that the microstructure may be 'released' and thus achieve mechanical movement. While the basic fabrication processes have been described for a silicon-based technology, there is much scope for a combination of other materials to act as a substrate for the mechanical microstructures and to act as a sacrificial layer.

Silicon-based fabrication technology is often preferred as silicon is a material for which the associated processes and equipment are extremely well understood, cheap and reliable. Furthermore, silicon MEMS processes are compatible with the complementary-metal-oxide semiconductor (CMOS) processes required to fabricate the signal-processing electronics. This has led to the design concept of integrated devices.

The main performance parameters required to estimate cost are the capital cost of the equipment, along with maintenance and other costs, and the time taken to perform an individual process, that is, the throughput of the machine. From the capital cost various parameters, such as depreciation, can be calculated. The expense of supporting the equipment can be calculated from the throughput and operator costs.

The basic design of common MEMS equipment and the associated processes will be outlined sufficiently for throughput and hence costs to be calculated. Routine and fault maintenance costs can vary from one machine to another and be directly attributed to each machine, if known. Alternatively, a fixed percentage can be attributed based on the capital cost of the machine, for example, 2% per annum.

2.2 Lithography, Masks and Reticles

Lithography is the process whereby an image of the given microstructure layer is projected onto a wafer surface, which has been pre-coated with a light-sensitive material known as a resist. After development of the resist, the wafer surface is either exposed or protected by the resist. The exposed area is then treated to further process steps such as material deposition or etching. Resists can be positive acting, where the underlying surface is exposed, or negative acting, where a defined area is covered or 'masked' (Mack, 2005).

Figure 2.1 shows a positive-acting resist, that is, one where the action of light and subsequent development selectively removes the resist and leaves the un-developed resist as a barrier to further process activity. Figure 2.2 shows a negative-acting resist, that is, one where the action of light and subsequent development leaves the resist as a barrier to further process activity and removes the un-developed resist. The resist materials are sensitive to ambient light,

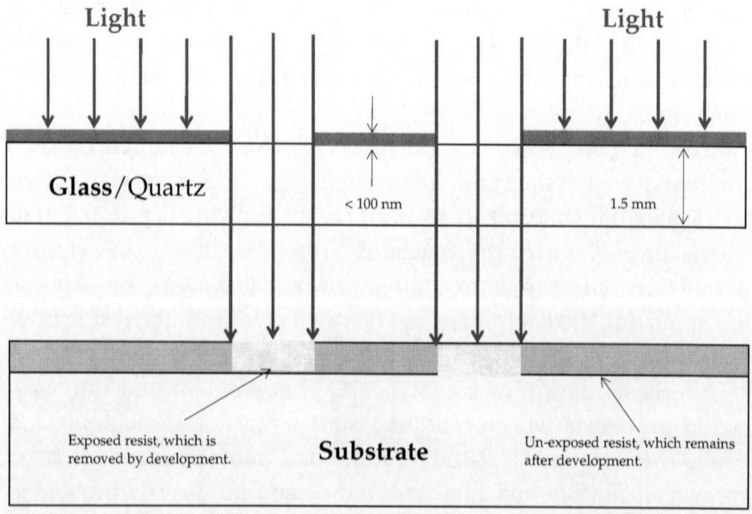

Figure 2.1 Optical chrome on quartz mask plates—positive-acting resist.

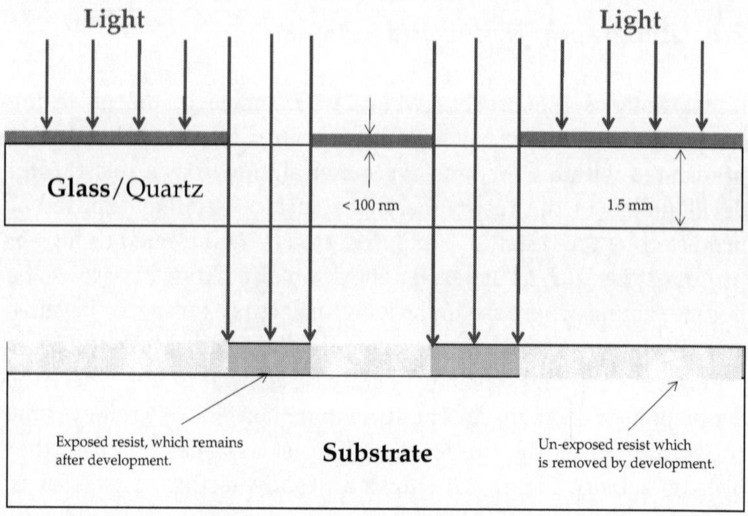

Figure 2.2 Optical chrome on quartz mask plates—negative-acting resist.

Figure 2.3 Layout of a wafer aligner mask.

so processing must be done in so-called 'yellow' rooms, where the wavelength of the ambient light is outside the sensitivity range of the resists (Microchem, 2011).

The process of lithography, resist exposure and processing, deposition and/or etching is repeated to fabricate each layer of the microsystem, thus building up the three-dimensional structure of the device. The initial component in this sequence is the plate holding the patterns to be used for the lithographic step. These plates are known as masks and/or reticles, depending on the type of optical projector used to pattern the resist.

A schematic of a mask is shown in Fig. 2.3 and consists of a thin (less than 100 nm) chromium layer deposited onto a thick (2.3 mm–6.35 mm) borosilicate or quartz substrate. The chromium is patterned and projected optically so that the dark/light structure is reproduced in the resist. There are two types of masks, the choice depending on the machine used to project the optical image.

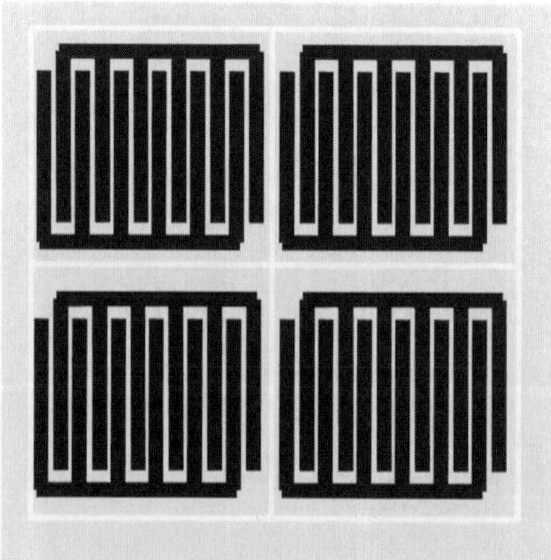

Figure 2.4 A reticle for a wafer stepper. Note, in this example, there are four dies per exposure.

In a wafer aligner (see Chapter 3, Section 3.2 for a full description), an image of the mask is projected across the whole area of the wafer. To expose multiple dies in one step the mask has images repeated on the mask itself (see Fig. 2.3) and the optical imaging is performed at unity (1:1) demagnification. A wafer aligner mask is usually referred to simply as a 'mask'.

A wafer stepper (see Chapter 3, Section 3.3 for a full description) produces a reduced image of the mask, now referred to as a 'reticle', which contains one or a few magnified images of the desired pattern. The stepper optics demagnify the reticle image, usually by 4:1 or 5:1, and this reduced reticle image, now die sized, is step-and-repeated across the wafer by the stepper itself (see Fig. 2.4).

Masks are usually manufactured in a commercial mask house and are readily available on the open market (Eynon & Wu, 2005). Mask and reticles for MEMS fabrication have a lower specification (e.g., 1 μm resolution) compared with masks for application-specific integrated circuit (ASIC) fabrication (e.g., <0.1 μm resolution) and hence are much cheaper (e.g., $1,000 per mask compared with $50,000 per reticle). In addition the number of masks/reticles required for MEMS is usually less than that for an ASIC process, for example, 8–14 compared with 20–30 reticles.

2.3 Deposition and Etching

The three-dimensional structure of a MEMS device is fabricated from a succession of material deposition and etching processes. Silicon-based MEMS processes use the same basic materials as the semiconductor industry, such as silicon, polysilicon, silicon nitride and silicon oxide. Polysilicon offers electrically conducting and mechanically robust microstructures and has SiO_2 as a sacrificial layer. Alternatively, metals can be used as mechanical microstructures, as well as aluminium for electrical interconnects. Detailed patterning is by the lithography procedures described in the previous section. The basic sequences for deposition and etching are shown in Fig. 2.5 and Fig. 2.6, respectively.

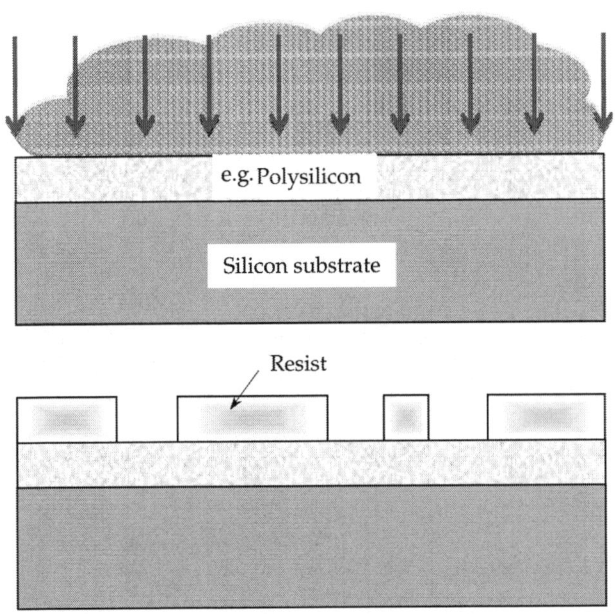

Figure 2.5 The fabrication processes of deposition and a patterned resist.

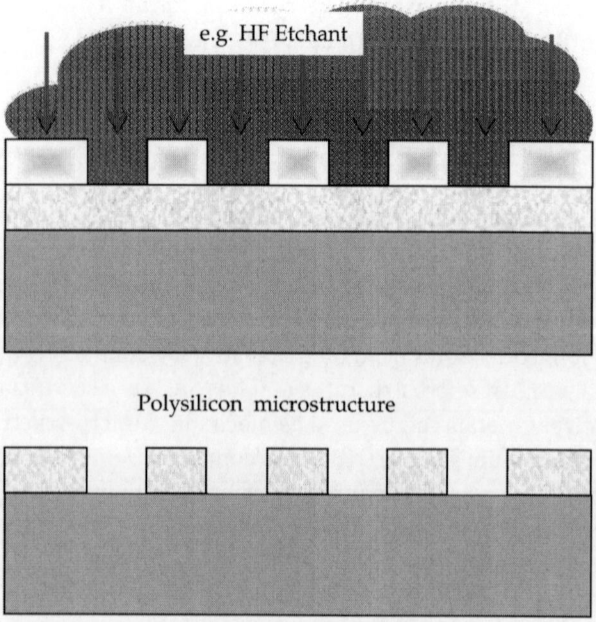

Figure 2.6 The fabrication processes of etching through a patterned resist.

In Fig. 2.5, the schematic assumes that a resist pattern has been formed and a layer of polysilicon deposited on the silicon substrate. Note that both materials are electrically conductive, and so a submicron layer of silicon oxide and silicon nitride may be required to electrically isolate the microstructure from the substrate (not shown).

In Fig. 2.6 the schematic shows the patterned resist on top of the polysilicon layer, followed by the etchant reaching the exposed polysilicon. The resist is then removed to reveal the desired microstructure. The equipment and processes commonly used to perform material deposition and etching are described in Chapter 3.

2.4 The Sacrificial Layer

A major difference between integrated circuits (ICs) and MEMS is that MEMS devices contain moving parts and hence require one or more special processes to fabricate a device on a micro scale. A major development is the 'sacrificial layer'.

MEMS fabrication forms the functional part of the mechanical component in a suitable material and uses a sacrificial material to temporarily support the microstructure. The sacrificial layer is subsequently removed to enable the microstructure to be freestanding. Figure 2.7 shows the basic steps necessary to form a

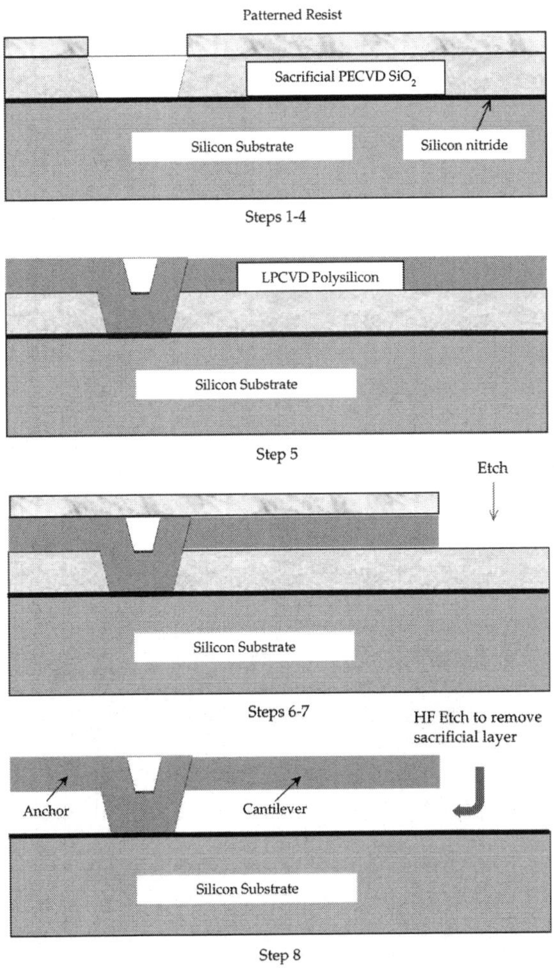

Figure 2.7 A sacrificial layer process for silicon MEMS.

moving part in silicon (a cantilever), where the functional material is polysilicon and the sacrificial layer is SiO_2, which is etched away by buffered hydrofluoric acid (HF).

The process starts with a silicon wafer coated with a sub-micron layer of silicon nitride to provide a dielectric surface on which the cantilever will rest.

The first process steps (steps 1–4) are to deposit the SiO_2 sacrificial layer, using plasma-enhanced chemical vapour deposition (PECVD), to pattern a window via the lithography resist-exposure process and to etch the exposed SiO_2. This will form a small cavity that will eventually anchor the moving part to the silicon surface.

The next step (step 5) is to deposit polysilicon using a low-pressure chemical vapour deposition (LPCVD) process, described in Chapter 3, Section 3.4. The polysilicon is then patterned by a lithographic sequence (step 6) that defines the length and width of the required cantilever. Step 7 is to create the cantilever by etching away the exposed polysilicon. The final step (step 8) is to remove the sacrificial SiO_2 to leave the polysilicon cantilever freestanding and anchored firmly to the silicon base.

A selection of the functional layers, the sacrificial layers and the etchants that can be used are shown in Table 2.1.

Table 2.1 Functional and sacrificial layers and etchant

Functional layer	Sacrificial layer	Etchants
Polysilicon	SiO_2	HF
Polysilicon	PSG	HF
Silicon nitride	SiO_2	HF
Silicon carbide	SiO_2	HF
Silicon nitride	Polysilicon	KOH
SiO_2	Polysilicon	TMAH
Aluminium	Silicon oxide	HF
Aluminium	Resist (AZ5214e)	O_2 plasma
Gold	Resist	O_2 plasma

Abbreviation: PSG, phosphosilicate glass.

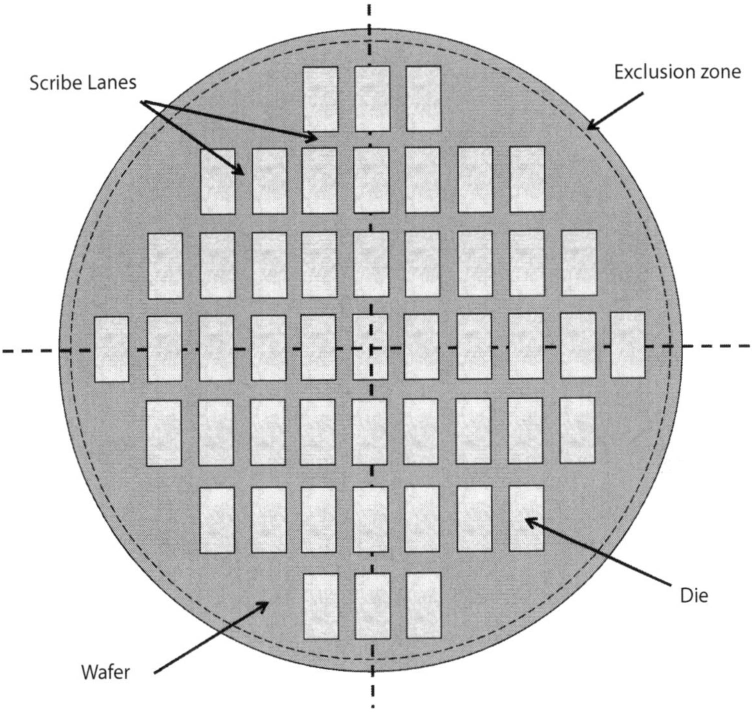

Figure 2.8 Layout of a step-and-repeated die on a wafer.

2.5 Number of Dies per Wafer

A schematic of step-and-repeated dies on a circular wafer is shown in Fig. 2.8. Wafer diameters are typically 200 mm and 300 mm for the semiconductor industry and 100 mm, 150 mm and 200 mm for the MEMS industry. Note the exclusion zone (typically 2–5 mm in width) around the perimeter of the wafer, which is required for equipment to load/unload the wafer and which is not available for die fabrication. The dies are separated from one another by 'scribe lanes' (also called 'streets'), typically 50–100 μm in width, which are used at the end of the fabrication process to allow diamond or laser cutting of the wafer into individual dies. The width of cut of the dicing saw is known as the 'kerf'.

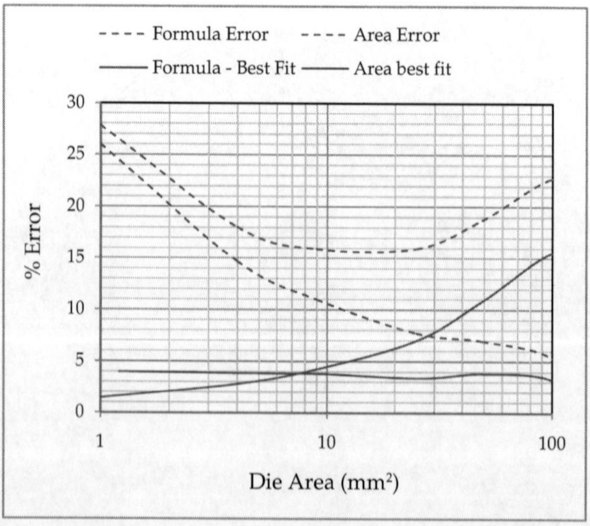

Figure 2.9 Error between the number of dies calculated by MEMSCOST (for a 200 mm wafer) and that calculated by the formula and the simple area method.

The number of dies per wafer will depend on the dimensions of the dies, the scribe lanes, the exclusion zone and the wafer size. Errors in calculating the number of dies per wafer will reflect on the final cost of the device, so it is important to use a calculation that is sufficiently accurate. The MEMSCOST die-per-wafer calculator takes into account not only the wafer and die dimensions but also the exclusion zone and scribe lanes. Options include square or rectangular dies and whether the central die is at the geometric centre of the wafer (as shown in Fig. 2.8) or whether the die is offset from the centre but symmetrical in the vertical or horizontal axis.

A common formula used in the industry for calculating the dies per wafer (N) is:

$$N = \pi D^2/4A_{\mathrm{D}} - \pi D/\sqrt{2A_{\mathrm{D}}} \tag{2.1}$$

where:

N = number of dies per wafer
D = wafer diameter
A_{D} = die area

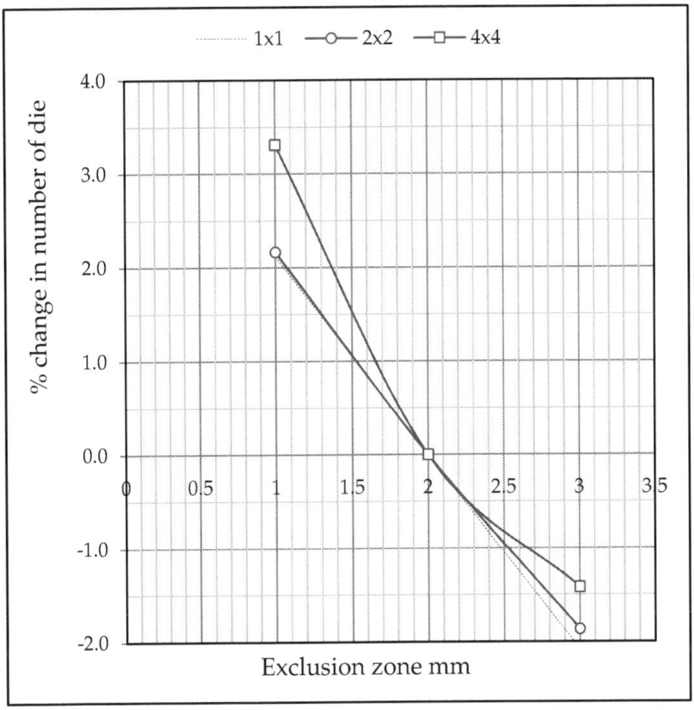

Figure 2.10 The effect of exclusion zone dimensions on dies per wafer.

This gives an approximation to the correct value but only for square dies. There are more accurate algorithms (Hackerott, 2004), including MEMSCOST, which gives values for both square and rectangular dies and the exact symmetry of the dies about the centre of the wafer. More importantly the die area, scribe lanes and exclusion zone are specified separately.

Simplistic models, based solely on dividing the wafer area by the die area, will produce significant errors in the final cost (in the region 3–4%). Figure 2.9 shows the percentage error made by the formula in calculating the number of square dies per wafer, as a function of die area, in comparison with MEMSCOST. The simple formula will give more accurate results if the die area used includes the scribe lane and the wafer diameter excludes the exclusion zone.

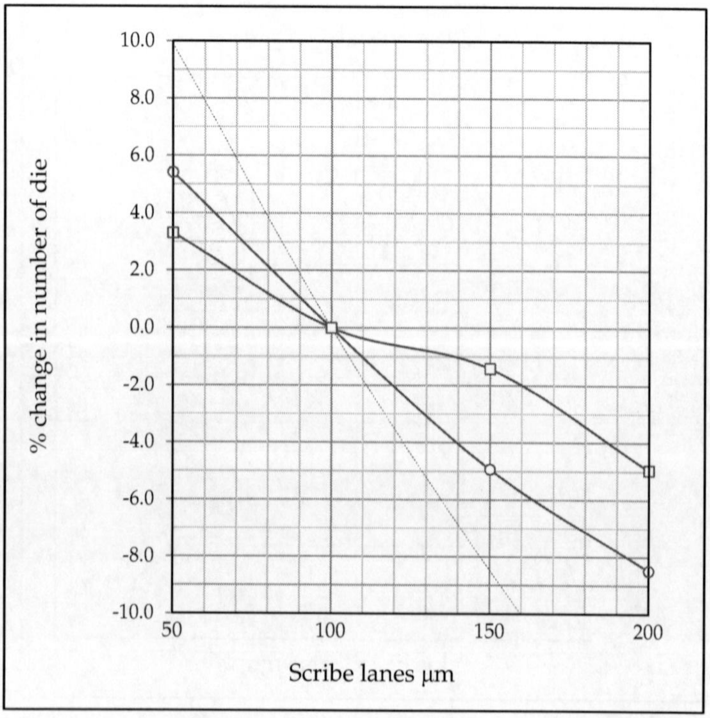

Figure 2.11 The effect of scribe lane dimensions on dies per wafer.

The dimensions of the exclusion zone and scribe lanes are determined by the requirements of the various machines involved in wafer handling and dicing for packaging. For the purposes of all calculations, these dimensions are fixed at 2 mm and 100 μm, respectively. However, variation of these parameters can have a significant effect on the number of dies that can be fitted on a wafer and therefore the unit cost of a device. Figure 2.10 estimates the percentage effect on the number of dies per wafer, and hence on cost, of varying the size of the exclusion zone. Figure 2.11 estimates the percentage effect on cost of varying the size of the scribe lanes.

The increased cost of manufacturing larger wafers is mitigated through having more die sites per wafer. Figure 2.12 shows the number of die sites per wafer diameter as a function of the die area (assuming a 2 mm exclusion zone and 100 μm scribe lanes). The

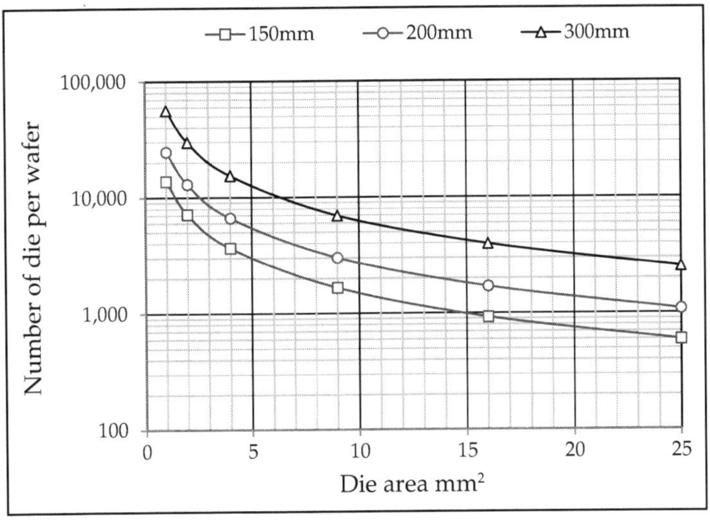

Figure 2.12 Dies per wafer vs. die area for different wafer diameters.

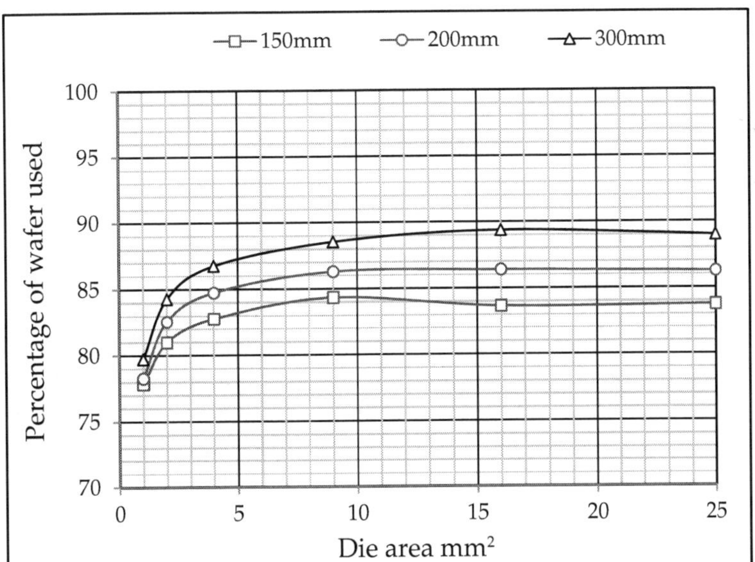

Figure 2.13 Percentage of wafer area used as a function of the die area for different wafer diameters.

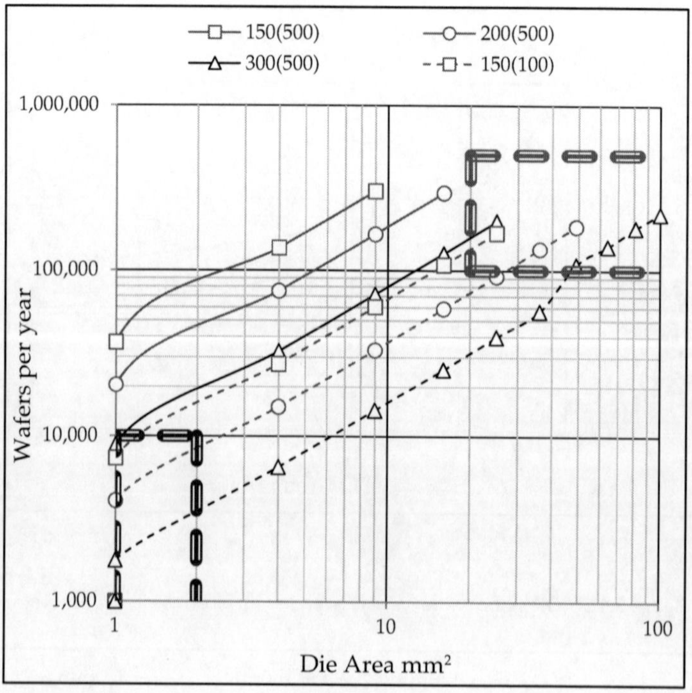

Figure 2.14 The wafers/year output required to meet a production quota of 100 million and 500 million devices as a function of the die area for 150 mm, 200 mm and 300 mm wafer diameters.

effect of an imperfect fit of a number of square dies to the available wafer area is shown in Fig. 2.13. Note that the benefit of fitting smaller dies to a given wafer diameter is reduced due to the need for more scribe lanes and that the larger wafers use the available area slightly more efficiently. Overall, it is difficult to use much more than 90% of the wafer area for any die area and wafer diameter combination.

In general, independent device manufacturers (IDMs) and foundries manufacturing ICs produce die sizes that are relatively large (e.g., 100 mm^2), and hence a large output of wafers per year is required to meet a demand of many millions of devices per year per factory (e.g., 1,000,000 wafers per year).

Figure 2.15 Part of a typical clean room containing fabrication equipment (courtesy of the Southampton Nanofabrication Centre).

The MEMS industry shows different characteristics, partly because the industry is less mature than the IC industry and is using smaller wafer sizes, partly because the devices tend to differ significantly in construction and materials and partly because the economics of fabrication are driving die sizes down below 1 mm^2.

Figure 2.14 shows the annual wafer output to deliver 100 million and 500 million devices, for a given die area, using 150 mm, 200 mm and 300 mm diameter wafers. The data is normalised to a yield of 100%. With a 200 mm wafer line and die areas of approximately 1 mm^2, the number of wafers is surprisingly low, for example, <10,000/year or 30/day, compared with the semiconductor industry, where a single 200 mm foundry can produce over 500,000 wafers per year. Also shown is the area bound by wafers per year and die area for MEMS and ICs.

In calculating the cost of devices, it is important that the loss of wafer area from the exclusion zone and scribe lanes be taken into consideration. Account must be taken of rectangular as well as square dies. MEMSCOST has an accurate model that automatically calculates the number of untested die sites per wafer.

2.6 Clean Rooms

Important parts of any microfabrication facility are the clean rooms, in which the equipment is housed. These are built to ultra-clean standards, with controlled temperature and humidity and with sub-micron particulates excluded (see Chapter 7). They are expensive to construct and operate. A typical MEMS clean room is shown in Fig. 2.15. Note that such a clean room would normally be yellow as optical resists are sensitive to normal light.

References

Eynon, BG, and Wu, B (2005). *Photomask Fabrication Technology.* McGraw Engineering Series.

Hackerott, M (2003). Die per wafer. mrhackerott.org/semiconductor-informatics/informatics/

Lawes, RA (1996). Fabrication technology for microengineering. *Sensor Rev.*, **16**(2), 16–22.

Levinson, HJ (2011). *Principles of Lithography, Third Edition. SPIE Monograph*, Vol. PM198.

Mack, C (2005). Fundamental *Principles of Optical Lithography: The Science of Microfabrication.* John Wiley.

Microchemical (2010). Exposure of photoresists. www.microchemicals.eu/technical_information.

Wolf, S, Tauber, RN (1986). *Silicon Processing for the VLSI Era—Volume 1 Process Technology.* Lattice Press. www.latticepress.com

Wolf, S, and Tauber, RN (1990). *Silicon Processing for the VLSI Era—Volume 2 Process Integration.* Lattice Press. www.latticepress.com

Chapter 3

Surface Micromachining

3.1 Introduction

Surface microengineering (Linder et al., 1992) is similar to conventional semiconductor fabrication in that a series of thin films are deposited, lithographically patterned and selectively etched. In the case of microelectromechanical systems (MEMS), a typical thin-film process is to deposit polysilicon layers (1–5 μm) onto a silicon dioxide sacrificial layer (1–5 μm). The silicon dioxide layer can then be selectively removed to leave a freestanding structure. Commercial foundry processes exist with up to five layers of polysilicon (see Chapter 11). Devices manufactured exclusively with surface microengineering are often referred to as 'complementary-metal-oxide semiconductor (CMOS)-MEMS'. Surface-microengineered devices are limited to near-planar options where the requirement is to detect extremely low capacitance changes (e.g., 10^{-15}–10^{-18} F) in an MEMS accelerometer or gyro. Compared with a surface-microengineered version, a high-aspect-ratio (HAR) sensor fabricated with deep microstructures will offer much higher signal levels (in proportion to their relative depth) and hence will require less demanding design.

MEMS Cost Analysis: From Laboratory to Industry
Ron Lawes
Copyright © 2014 Pan Stanford Publishing Pte. Ltd.
ISBN 978-981-4411-06-6 (Hardcover), 978-981-4411-07-3 (eBook)
www.panstanford.com

The processes described in this chapter do not include all of those used in CMOS and the hybrid bipolar transistor CMOS (BiCMOS) manufacture, but a selection is offered that are also employed in MEMS fabrication. However, simple models for a notional CMOS/BiCMOS process are included in MEMSCOST (see Chapter 8) so that estimates can be made for MEMS devices that include the signal-processing microelectronics.

3.2 Optical Mask Aligner

The optical aligner, or wafer aligner, was an early lithography machine where the minimum dimensions required for semiconductor manufacture were approximately 1.5 μm or greater. A mask depicting a full-size die is imaged directly onto the wafer and exposed and the exposed wafer unloaded from the machine and then replaced by the next unexposed wafer. Alignment of the new die image to any previously processed microstructure on the wafer is undertaken manually by using an optical alignment system capable of an accuracy of 0.15 μm.

A high-pressure mercury lamp provides exposure illumination up to 350 W integrated over the useful ultraviolet (UV) wavelengths. The spectrum of a mercury lamp is shown in Fig. 3.1 illustrating the domination of the g-line (436 nm), h-line (405 nm) and i-line (365 nm) wavelengths. The resolution (R) of aligner optics is limited by Fresnel diffraction (Gerloch, 2008) and is approximately

$$R = 1.5\sqrt{\lambda(s + t/2)} \tag{3.1}$$

where:

s = gap between mask and wafer surface
t = resist thickness and $t/2$ allowing for light scattering in the resist itself
λ = wavelength of the illumination

From Eq. 3.1, it will be seen that resolution is a function of the proximity of the mask to the wafer, and hence a wafer aligner is often referred to as a 'proximity printer'. As the minimum

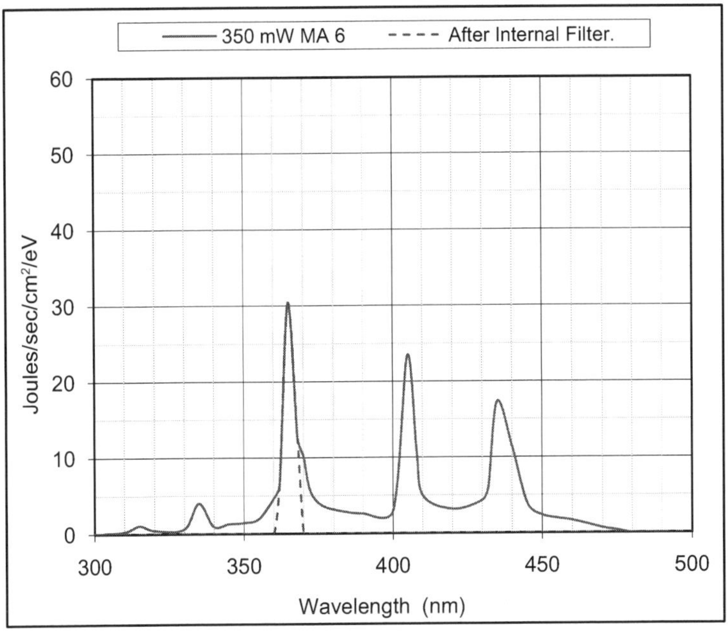

Figure 3.1 Energy spectrum of a mercury lamp.

dimensions required by semiconductor technology reduced according to Moore's law (Hutcheson, 2005), the wafer aligner could deliver neither the resolution nor the alignment accuracy required for deep sub-micron lithography. Consequently, the wafer aligner has been superseded, for semiconductor applications, by the optical wafer stepper (see Section 3.3). Nevertheless, the wafer aligner is a cost-effective printer for MEMS lithography. A MEMS mask for an aligner does not require high resolution and typically can be purchased for $1,000 per mask.

Resolution with a wafer aligner is often good enough to fabricate MEMS devices, with the mask-to-wafer gap sufficiently large to avoid damaging delicate parts. For example, from Fig. 3.1, there is a strong intensity at a wavelength of 365 nm, which, from Fig. 3.2, would enable MEMS features down to a 3.5 μm resolution to be replicated in 10 μm of a resist with a gap of approximately 10 μm, sufficient to

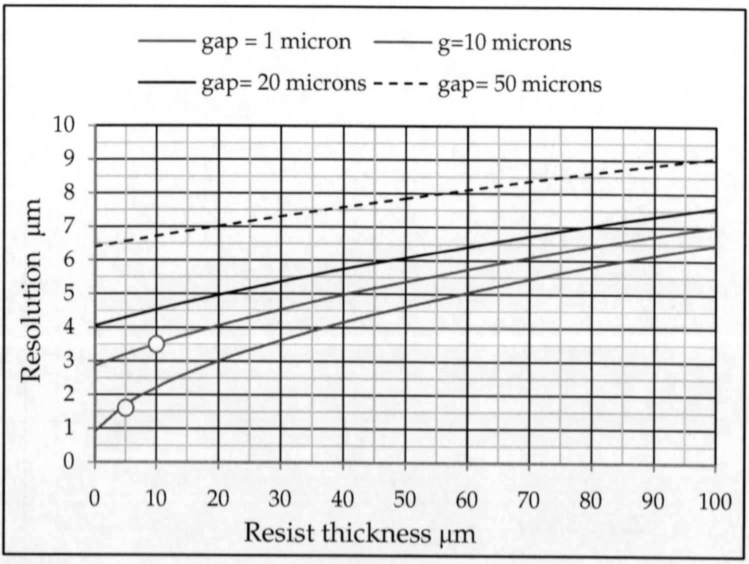

Figure 3.2 Wafer aligner resolution with a 365 nm wavelenth source as a function of the mask-wafer gap and resist thickness. Resolution for a 10 μm gap and a 10 μm resist is shown at 3.5 μm and 1.5 μm, respectively.

protect vulnerable microstructures. Resolution down to 1–1.5 μm can be achieved if the gap is reduced to virtually mask-to-wafer contact.

A schematic of a typical mask aligner is shown in Fig. 3.3 and a commercial machine in Fig. 3.4. The wafer is coated with an optical resist, for example, AZ9260 (Microchem, 2011), which has a sensitivity of 500 mJ/cm^2. The exposure time would be 20 seconds, which, allowing for a wafer handling and registration overhead of 35 seconds, results in a throughput of 45 wafers/hour. An estimate of the wafer aligner throughput is shown in Table 3.1.

The ability to align to the bottom side of the wafer, referred to as 'double-side alignment', has been added to the original design of wafer aligners, thus allowing more flexibility in MEMS design. One implementation (SUSS, 2011) uses cross-hair alignment marks on the bottom side of the wafer viewed through a second objective lens on the wafer aligner.

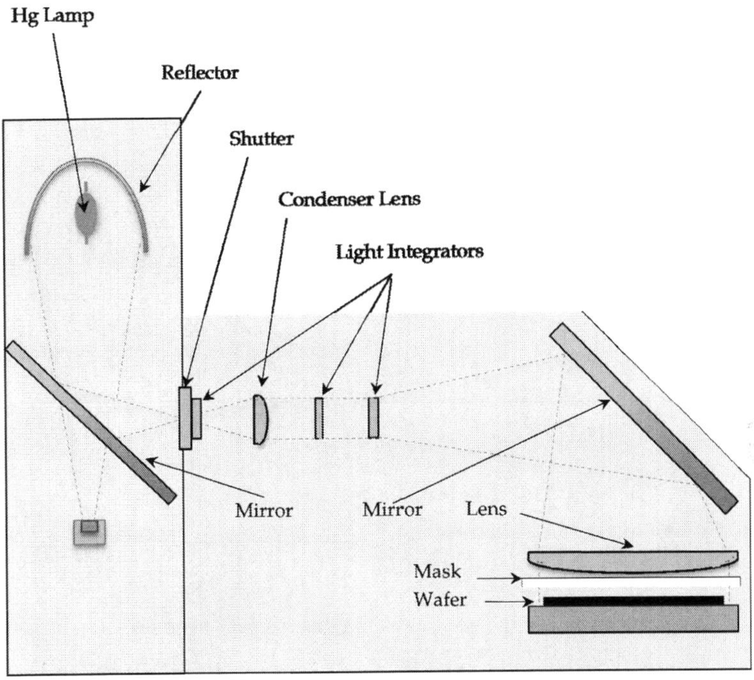

Figure 3.3 Schematic of an optical wafer aligner.

Table 3.1 Wafer aligner machine simulator

Parameter	Dimension
Lamp power (g-h-i line)	350 W
Power density	25 mW/cm^2
Optical resist sensitivity	500 mJ/cm^2
Exposure time	20 seconds
Load/unload (inc. 10 s alignment)	70 seconds
Total time	90 seconds
Batch size	1 wafer
Throughput	40 wafers/hour

Figure 3.4 EVG6200 optical mask aligner (courtesy of the Southampton Nanofabrication Centre).

3.3 Optical Wafer Stepper

The optical wafer stepper was developed to meet the growing demand for increased resolution for the integrated circuit industry, firstly sub-micron and then sub-100 nm. Currently, with developments in mask, illumination and resist processing, 22 nm is available. It will be seen that much earlier models of wafer steppers can deliver sub-micron resolution to MEMS fabrication, where required.

An optical wafer stepper images one or several dies at a time onto a wafer from a mask via a refractive lens. The mask used for an optical stepper is usually referred to as a reticle to distinguish it from the full wafer masks used with mask aligners. The reticle-to-wafer demagnification is typically 4–5, with a working lens diameter in the order of 30 cm. With reticle dimensions typically 150 mm $\times$ 150 mm, it is possible to image an array of dies in a single exposure, depending on the die and reticle size, thereby increasing throughput.

The wafer is held on a precision stage whose position, and that of the wafer, can be digitised to a few nanometers by laser interferometers. The stepper provides exceptional resolution and alignment accuracy (the positioning of the image on one layer of a die with respect to a die on a previous exposure). The stepper aligns the image to be exposed on special alignment marks to an accuracy of approximately $+/-10$ nm and then exposes the new image. The stage then moves the position of the wafer with respect to the optical system to a new die position, and alignment takes place, followed by resist exposure. The process is repeated until all dies on a particular layer of a wafer are exposed. The process of repeated movement of the stage, alignment and resist exposure is known as 'step-and-repeat'.

The resolution of a stepper lens (R) can be calculated from Eq. 3.2 and is shown graphically in Fig. 3.5. The resolution and typical numerical aperture are also shown for the i-line and laser source, where, to reduce resolution, the mercury lamp source has been replaced with an excimer laser that has a shorter wavelength (λ). Typical laser wavelengths are 248 nm (krypton fluoride) and 193 nm (argon fluoride). The resolution and depth of focus (DoF) are given by Eq. 3.2 and Eq. 3.3, respectively.

$$R = k_1\lambda/\text{NA} \tag{3.2}$$

$$\text{DoF} = k_2\lambda/\text{NA}^2 \tag{3.3}$$

where:

k_1, k_2 = constants depending on the optics of the stepper and including the mask structure

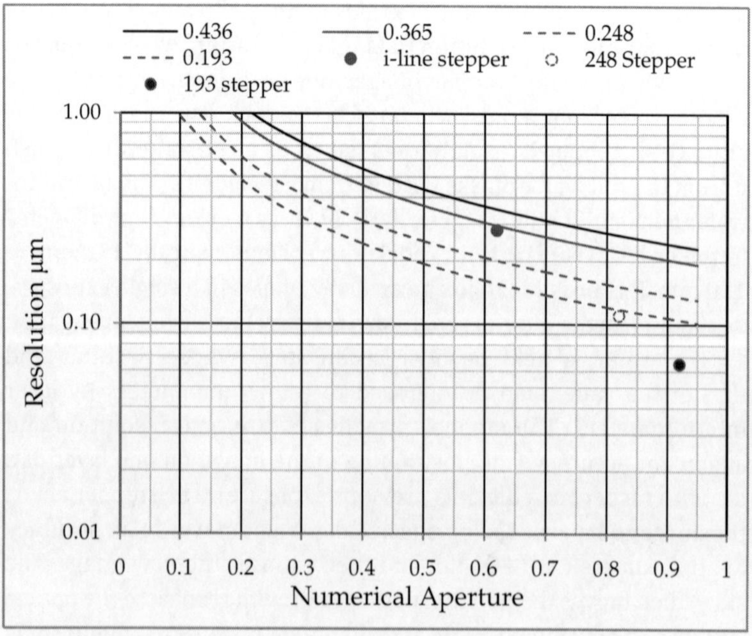

Figure 3.5 Resolution as a function of the numerical aperture of the projection lens. The results are derived from Eq. 3.2 for the parameters shown.

NA = numerical aperture of the lens
λ = wavelength of the illumination system

For deep sub-micron resolution, several novel optical and resist-processing methods have been developed to reduce k_1 (Eq. 3.2), such as phase shift masks, off-axis illumination and optical proximity correction (Levinson, 2011).

However, it will be seen from Eq. 3.3 that increasing the numerical aperture to improve resolution reduces the DoF, that is, the extent to which the lens accurately images below the focal plane (which is usually the resist surface). This effect has placed emphasis on planar surfaces and the introduction of techniques such as chemical mechanical polishing (CMP) (see Section 3.10).

A schematic of a basic step-and-repeat stepper is shown in Fig. 3.6 and a typical machine in Fig. 3.7. The optical wafer stepper is the production tool of choice for the semiconductor industry, and

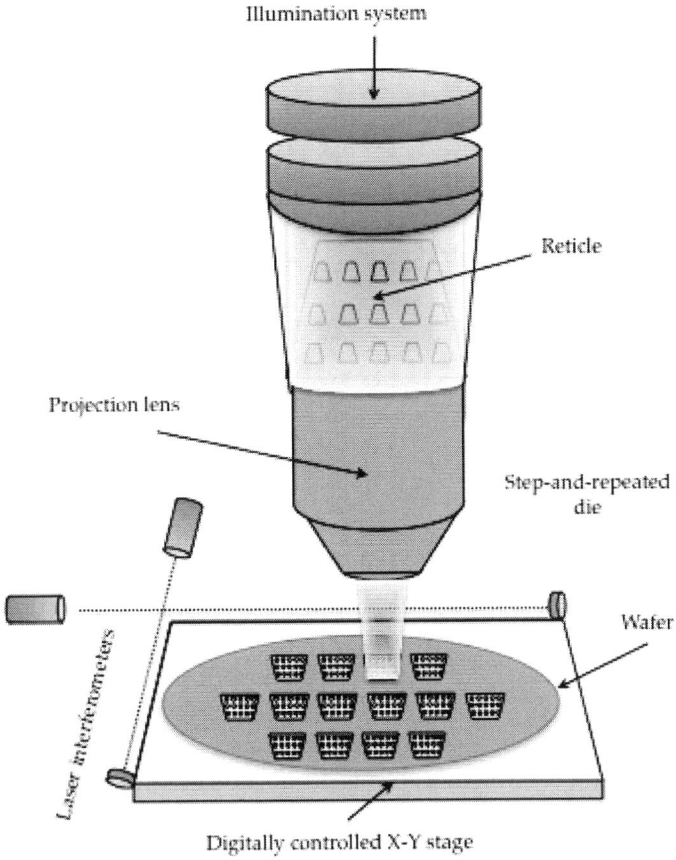

Figure 3.6 Schematic of an optical wafer stepper.

the equipment suppliers have developed steppers with increasingly advanced techniques, for example, by combining a step-and-repeat strategy with simultaneous scanning of the reticle and stage.

The optical wafer stepper is used for MEMS when design rules in the order of a micron need to be implemented or when second-hand steppers are available as part of a converted CMOS line. A commonly available machine is the i-line stepper, which, from Eq. 3.2, will typically produce a 0.4 μm resolution ($k_1 = 0.5$, $\lambda = 0.365$ nm, NA $=$ 0.5). Manufacturers of steppers quote an even better resolution by

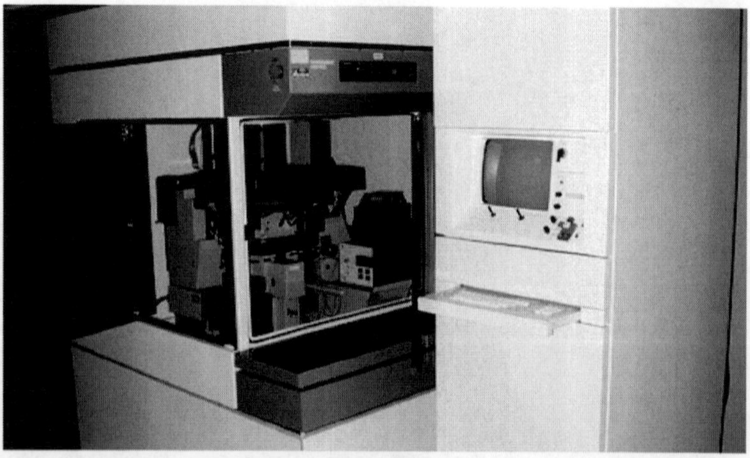

Figure 3.7 Optical wafer stepper (courtesy Qudos Ltd.).

using k values in the region of 0.35–0.4 compared with 0.5 (ASML, 2012; Canon, 2012; Nikon, 2012).

Use of a stepper will affect costs as throughput depends on interlinked factors such as wafer size, lens diameter, die size, resist exposure time with a given illumination source and the total time taken for the stage to step to a new position and then align and expose the die image. In the example shown in Table 3.2, the lens projects an image of a 10 × 10 array of dies requiring 78 ms to move the stage to a new position, 250 ms to align and 200 ms to expose the resist. This has to be repeated 100 times, requiring 32.8 seconds to move and align and 53 seconds to expose the wafer. With a small allowance to load/unload wafers from a multi-wafer cassette, the exposure time per wafer is 86.5 seconds, equivalent to 41.6 wafers per hour.

Mask/reticle prices from a commercial mask shop are expensive for CMOS, as shown in Fig. 3.8. For example, for a 0.13 μm, 26-mask-set process, the estimated cost would be $650,000 ($25,000 per mask). A mask set for a MEMS process should be much cheaper, for example, a 0.8 μm, 16-mask-set process would cost $26,000 ($1625 per mask), that is, it would be more expensive than a wafer aligner mask.

Table 3.2 Optical wafer stepper simulator

Parameter		Dimension		
Wafer diameter	200	mm		
No. of dies	6,656			
Reticle magnification	5			
Lens diameter	30	mm		
Length	21.2	mm		
Die length	2	mm		
Die width	2	mm		
Die image/reticles (X)	10			
Die image/reticles (Y)	10			
Steps in X	10			
Steps in Y	10			
Total	100			
Acceleration	15,000	mm/second^2		
Velocity	500.0	mm/second		
Time/die L/V	0.073	seconds		
Settling time	0.005	seconds		
Registration	0.25	seconds		
Move time/image	0.328	seconds	Total move time	32.8
Power density	1,000	mW/cm^2		
Resist sensitivity	200	mJ/cm^2		
Exposure time	0.20	seconds		
Time/reticle	0.53	seconds	Time/wafer	53
Load/unload	0.25	(25 wafers/100 seconds)		
Wafer stage	0.5	seconds		
Load/unload	0.75	seconds	Total time	86.55
			Wafers/hour	41.6

3.4 Low-Pressure Chemical Vapour Deposition

LPCVD produces solid material through a chemical reaction in a furnace, which is deposited on the substrates present (Jaeger, 2002). For MEMS, LPCVD is normally used to deposit polysilicon and silicon nitride. Silicon dioxide can be deposited by LPCVD, but plasma-enhanced chemical vapour deposition (PECVD) is often preferred, as the reaction takes place at lower temperatures (see Section 3.5). A schematic of the basic LPCVD furnace arrangement is shown in Fig. 3.9 and a typical machine in Fig. 3.10. Various gases are introduced into the furnace and heated to the operating temperature for the

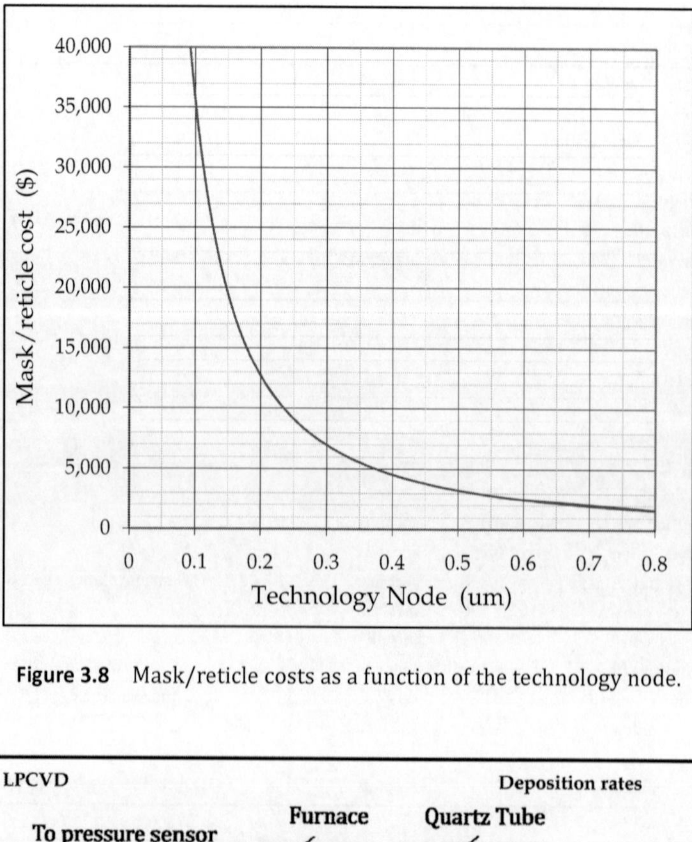

Figure 3.8 Mask/reticle costs as a function of the technology node.

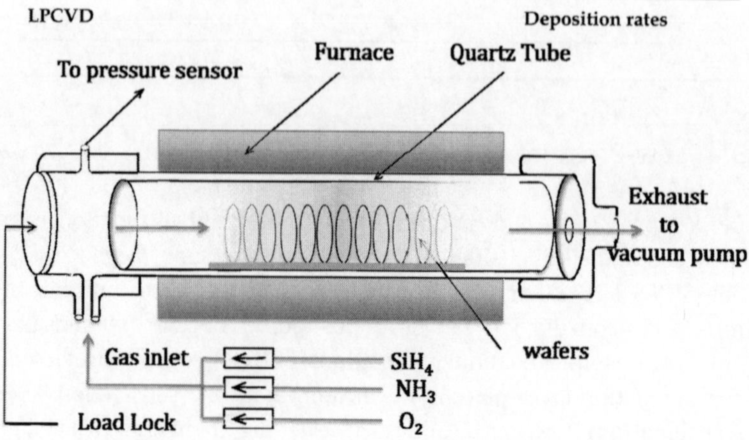

Figure 3.9 Schematic of a LPCVD furnace.

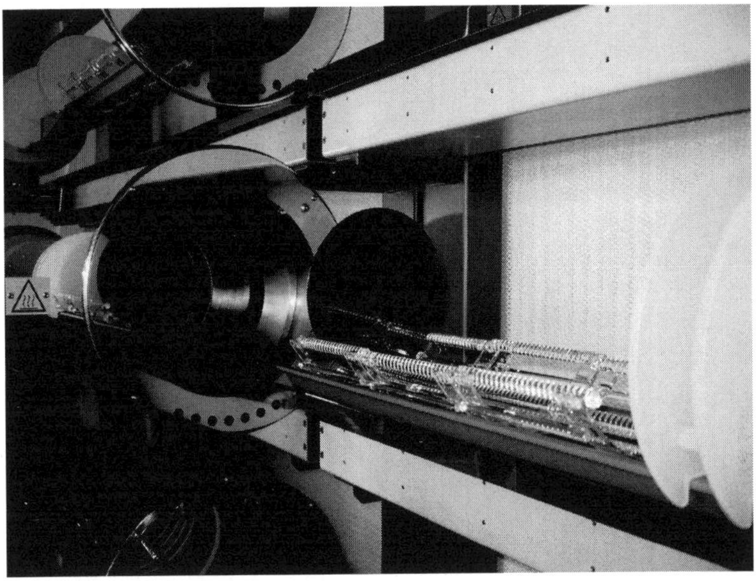

Figure 3.10 LPCVD furnace (courtesy of the Southampton Nanofabrication Centre).

necessary chemical reaction. The various by-products are pumped from the reactor and entered into a burn-off furnace for safety. The chemistry of the processes regularly used in LPCVD is outlined below.

For polysilicon deposition using silane at 600–700°C:

$$SiH_4 \rightarrow Si + 2H_2 \tag{3.4}$$

For silicon nitride deposition using silane and ammonia at 700–900°C (Eq. 3.5) or dichlorosilane + ammonia at 700–800°C (Eq. 3.6):

$$SiH_4 + 4NH_3 \rightarrow SiN_4 + 12H_2 \tag{3.5}$$

$$SiH_2Cl_2 + 4NH_3 \rightarrow SiN_4 + HCl + 6H_2 \tag{3.6}$$

For silicon dioxide deposition, several processes are available, such as tetraethyloxysilane at 650–750°C (Eqs. 3.7 and 3.8), silane and oxygen at 450°C (Eq. 3.9) or dichlorosilane + nitrous oxide at 900°C (Eq. 3.10).

$$Si(OC_2H_5)_4 + 2H_2O \rightarrow SiO_2 + 4C_2H_5OH \tag{3.7}$$

Table 3.3 LPCVD simulator

LPCVD			Deposition rates Material	nm/minute
Material	Polysilicon		Polysilicon	10
Rate	10	nm/minute	SiO_2	25
Thickness	1	micron	Si_3N_4	5
Time	100	minutes		
Load/unload time	1	minute		
Total time	101	minutes		
Batch/hour	0.6			
Wafers/batch	25			
Throughput	14.9	wafers/hour		

$$Si(OC_2H_5)_4 \rightarrow SiO_2 + 2O(C_2H_5)_2 \tag{3.8}$$

$$SiH_4 + O_2 \rightarrow SiO_2 + H_2 \tag{3.9}$$

$$SiH_2Cl_2 + 2N_2 \rightarrow SiO_2 + 2N_2 + 2HCl \tag{3.10}$$

Commonly used metals such as aluminum and copper normally use a sputtering technique (see Section 3.6), although some metals may be deposited by LPCVD, for example, tungsten at 300–500°C (see Eq. 3.11). The LPCVD machine simulator, used in MEMSCOST to calculate throughput, is shown in Table 3.3 for typical materials used in MEMS fabrication.

$$WF_6 + 3H_2 \rightarrow W + 6HF \tag{3.11}$$

3.5 Plasma-Enhanced Chemical Vapour Deposition

PECVD can be used to deposit low-tensile-stress silicon oxide for MEMS and, in some applications, polysilicon, silicon nitride and some metals (Jaeger, 2002). PECVD has the advantage over LPCVD of requiring much lower temperatures to deposit material at 300–350°C, compared with 600–900°C for LPCVD.

The basic furnace arrangement is shown in Fig. 3.11 and a typical machine in Fig. 3.12. Plasma is generated by a 13.56 MHz radio frequency (RF) source (e.g., at 1,000 V, 600 W), which energises

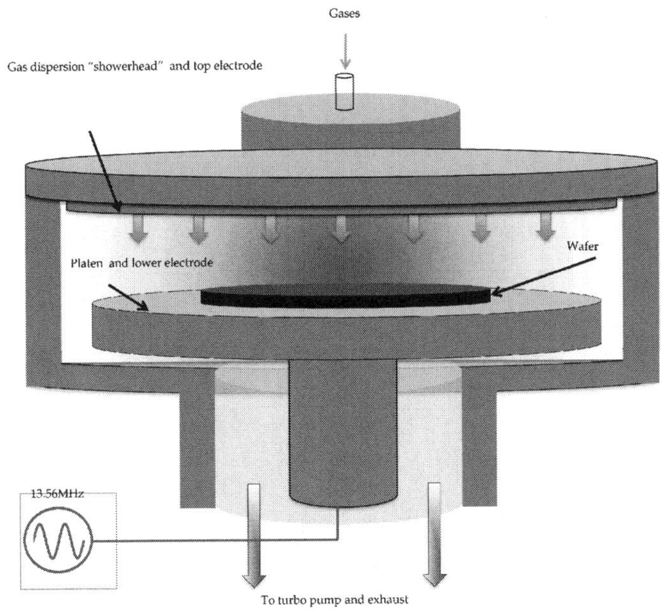

Figure 3.11 Schematic of a PECVD machine.

the gas mixture to produce the required chemical reaction. The gas mixtures and chemical reactions are the same as for LPCVD but at lower temperatures. The PECVD machine simulator, used in MEMSCOST to calculate throughput for typical materials used in MEMS fabrication, is shown in Table 3.4. In this example the throughput is high at 200 wafers per hour.

For polysilicon deposition using silane at 600–700°C:

$$SiH_4 \rightarrow Si + 2H_2 \tag{3.12}$$

For silicon nitride deposition using silane and ammonia (Eq. 3.13) or dichlorosilane + ammonia at 200–500°C (Eq. 3.14):

$$SiH_4 + 4NH_3 \rightarrow SiN_4 + 12H_2 \tag{3.13}$$

$$SiH_2Cl_2 + 4NH_3 \rightarrow SiN_4 + HCl + 6H_2 \tag{3.14}$$

For silicon dioxide deposition, several chemicals are available, such as tetraethyloxysilane at 300–500°C (Eqs. 3.15 and 3.16), silane

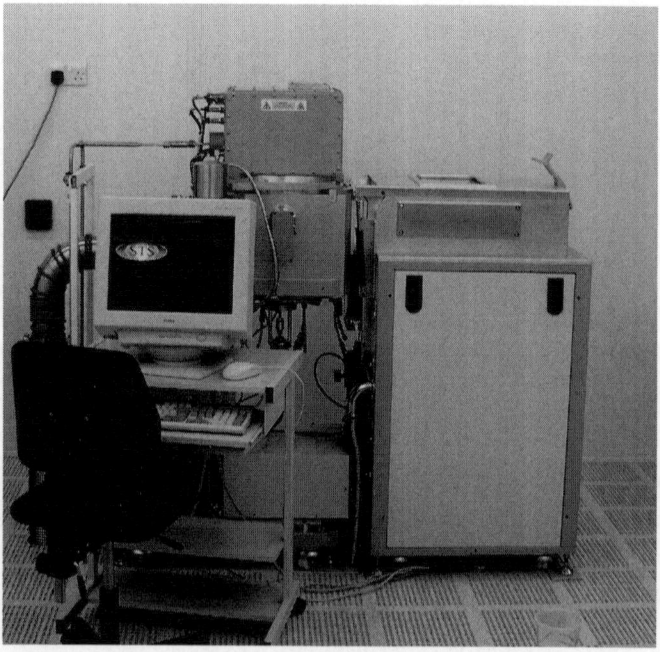

Figure 3.12 PECVD polydeposition machine (courtesy Scottish Microelectronics Centre).

and oxygen at 200–400°C (Eq. 3.17) or dichlorosilane + nitrous oxide at 200–400°C (Eq. 3.18).

$$Si(OC_2H_5)_4 + 2H_2O \rightarrow SiO_2 + 4C_2H_5OH \quad (3.15)$$

$$Si(OC_2H_5)_4 \rightarrow SiO_2 + 2O(C_2H_5)_2 \quad (3.16)$$

$$SiH_4 + O_2 \rightarrow SiO_2 + H_2 \quad (3.17)$$

$$SiH_2Cl_2 + 2N_2 \rightarrow SiO_2 + 2N_2 + 2HCl \quad (3.18)$$

3.6 Thermal Oxide Growth

Thermal oxide is an excellent insulator and can produce good conformal cover over non-planar microstructures (Sze, 2001).

Table 3.4 PECVD simulator

PECVD			Deposition rates Material	nm/minute
Material	SiO_2		SiO_2	200
Rate	200	nm/minute	Si_3N_4	100
Thickness	1	micron		
Time	5	minutes		
Load/unload time	1	minute		
Total time	6	minutes		
Batch/hour	10			
Wafers/batch	1			
Throughput	10	wafers/hour		

A disadvantage for its use elsewhere in MEMS fabrication is that high temperatures are involved (900–1,200°C) for thermal oxide deposition, whereas LPCVD and PECVD can produce the oxide at lower temperatures. A thermal oxide reactor is shown diagrammatically in Fig. 3.13 and a typical installation in Fig. 3.14. There is a wet and a dry process shown in Eq. 3.19 and Eq. 3.20, respectively. The wet process uses steam on silicon, and the dry process uses oxygen on silicon.

$$Si + 2H_2O \rightarrow SiO_2 + 2H \tag{3.19}$$

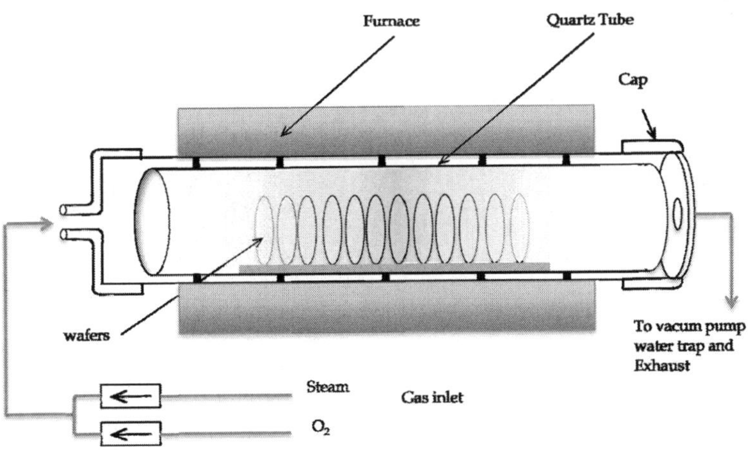

Figure 3.13 Thermal oxide reactor.

Figure 3.14 Thermal oxide reactor (courtesy Southampton Nanofabrication Centre).

$$Si + O_2 \rightarrow SiO_2 \tag{3.20}$$

The wet process is preferred to the dry process, as the deposition rate is much higher. The processes convert silicon to silicon dioxide, part of which is grown into the bare silicon surface (46%), with the

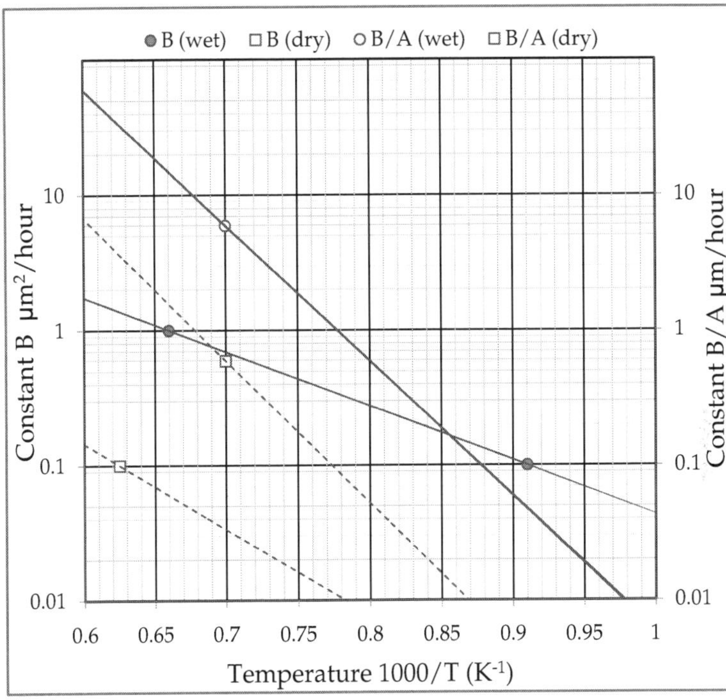

Figure 3.15 Thermal oxidation constants for wet and dry oxidation as a function of temperature in degrees Kelvin.

remainder (54%) grown as the required layer thickness above the silicon surface. The deposition rate is a function of temperature and slows down as the deposition time increases. The deposition time can be calculated from the Deal–Grove model (Deal & Grove, 1965; Grove, 1971) from Eqs. 3.21 and 3.22 and is shown in Fig. 3.15.

$$X_T^2 + AX_T = B(t + \tau) \tag{3.21}$$

Solving the quadratic for X_T

$$X_T = (A/2)\sqrt{1 + 4B(t + \tau)/A^2 - 1} \tag{3.22}$$

where:

X_T = total thickness of oxide above and below the silicon surface

t = time taken to grow an oxide thickness X_T

T = time that would have been required to grow any pre-existing oxide

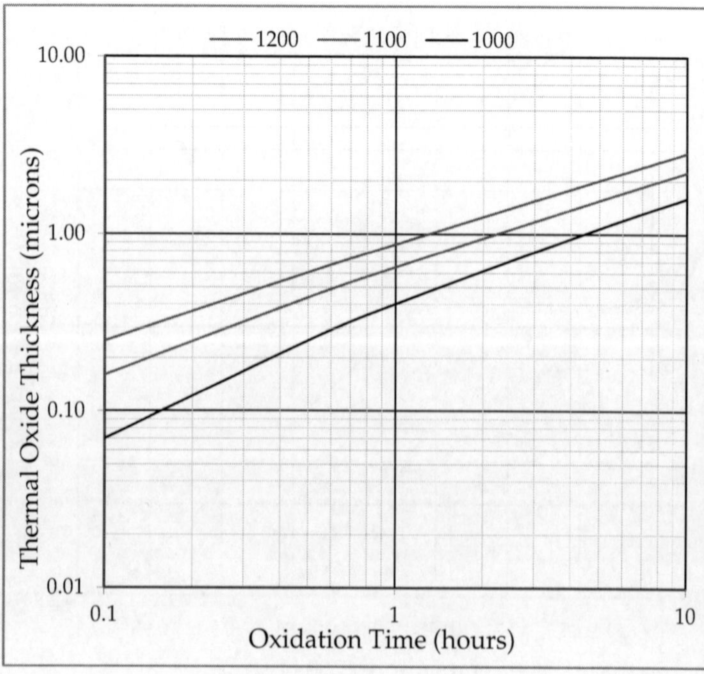

Figure 3.16 Thermal oxide deposition at 1,000°C, 1,100°C, and 1,200°C.

B = linear constant and a function of temperature
B/A = parabolic constant and a function of temperature

The constants B and B/A for wet and dry oxidation are shown in Fig. 3.15.

Figure 3.16 shows results for X_T, assuming no pre-existing oxide, and for temperatures varying from 1,000°C to 1,200°C. This is the basis of the MEMSCOST machine simulator shown in Table 3.5, which calculates the thickness of silicon dioxide that is deposited above the silicon surface. For example, a 0.650 μm layer above a bare silicon surface, at a temperature of 1,100°C, requires a deposition time of approximately one hour, an effective thin-film deposition rate of 10.8 nm per minute (note that the deposition rates decrease as the deposition time increases). A thermal oxide calculator is also available on the Internet (Brigham Young University, 2012).

Table 3.5 Thermal oxide simulator (<100> silicon substrate)

			Deposition rates	
Thermal oxide	Parameters		Temperature °C	nm/minute
Process	Wet			
Temperature °C	1,100		1,000	4
Thickness	1.000		1,100	2.1
Deposition time	2.14	hours	1,200	1.3
Load/unload	1	minute		
Total	2.16	hours		
Batch size	25			
Throughput	11.6	wafers/hour		

3.7 DC and RF Sputter Deposition

Sputtering by physical vapour deposition (PVD) enables a thin film of material to be deposited on wafers, whether they are processed or partially processed. The material is vapourised by plasma, thereby producing ions that bombard the target material. The resultant sputtered material then forms a film on the nearby substrate. The optimum deposition rate is achieved in a chamber pressure of approximately 100 mTorr.

To minimise particulate contaminants, the sputtering is usually done in a sputter-up mode, that is, with the sputtering target at the bottom of the chamber and the substrate at the top. Sputtering can be done with a direct current (DC) or RF source, depending on the material to be deposited. Configurations for DC and RF sputtering are shown in Fig. 3.17 and Fig. 3.18, respectively, and a typical DC machine in Fig. 3.19. Various additional components are normally available, such as substrate bias voltage and temperature, in order to improve film quality. The sputter deposition machine simulator used in MEMSCOST to calculate throughput is shown in Table 3.6 for typical materials used in MEMS fabrication.

Typically with DC sputtering argon gas is ionised and used to sputter material from a target made of the appropriate metal. DC sputtering is a widely used technique for depositing metal layers, such as aluminum, gold, platinum, tungsten and various metal alloys, in both semiconductor and MEMS fabrication.

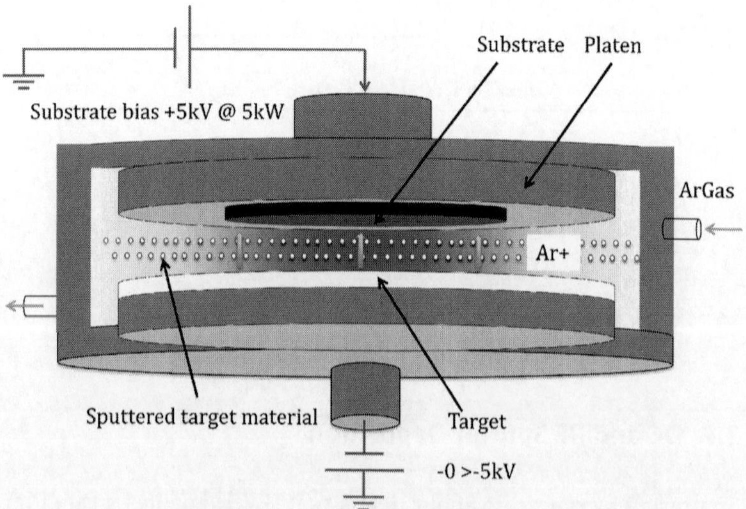

Figure 3.17 DC sputter machine.

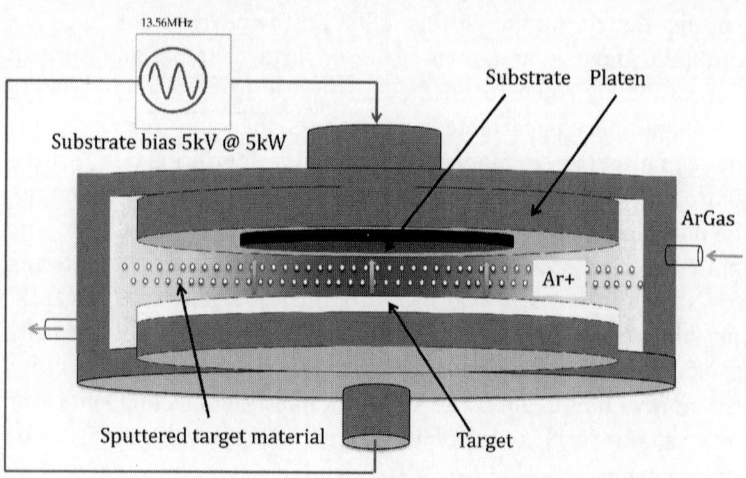

Figure 3.18 RF sputter machine.

Figure 3.19 DC metal sputter deposition machine (courtesy Scottish Microelectronics Centre).

The deposition rate will depend on the power levels applied to the metal target, the excitation voltage, the source-to-substrate distance, the background pressure and the sputter yield of the target material (number of atoms ejected from the target material per incident ion). Note that the sputter yield depends on the target material, the mass of the bombarding ion and its energy.

DC sputtering of dielectrics (e.g., SiO_2, Si_3N_4) is impractical due to charge build-up on the substrate. However, RF sputtering can be used, as electrons are available to neutralise charge build-up on the insulating substrate. Silicon dioxide and silicon nitride are typical

Table 3.6 DC (metal) and RF (dielectrics) simulator sputter rates for some typical materials

			Deposition rates	
Sputter deposition			Material	nm/minute
Material	Aluminium	SiO_2	14	
Rate	15	nm/minute	Si_3N_4	12
Thickness	1	micron	Polysilicon	13
Time	66.7	minutes	Aluminium	15
Load/unload time	1	minute	Titanium	5
Total time	67.7	minutes	Chromium	7
Batch/hour	1.13		Gold	16
Wafers/batch	1			
Throughput	1.13	wafers/hour		

materials that can be deposited. A 13.56 MHz RF generator provides the excitation of the argon gas (Bhatt et al., 2007; 2009).

Deposition rates can be increased by incorporating magnets behind the target, thereby creating a crossed magnetic and electric field and increasing the argon ionisation rate and hence the deposition rate. This technique is known as magnetron sputtering (Junghoon, 2000).

3.8 Wet Etching

Wet-chemical etching has been the most widely used technique for processing silicon since the earliest development of semiconductors and MEMS, as the procedures have been relatively simple, fast and cheap. All that is required is a suitable wet bench, with appropriate protection against corrosive chemicals, a supply of the chemical etchants and an etch mask material that either dissolves very slowly compared with the film/substrate or does not dissolve at all. Wet-chemical processing is particularly suitable for etching thin films or the substrate itself.

For silicon substrates, the etchant progresses through the material in two different ways, depending on the etchant and the crystallographic orientation of the silicon. Figure 3.20 shows the microstructures formed by isotropic and anisotropic etching.

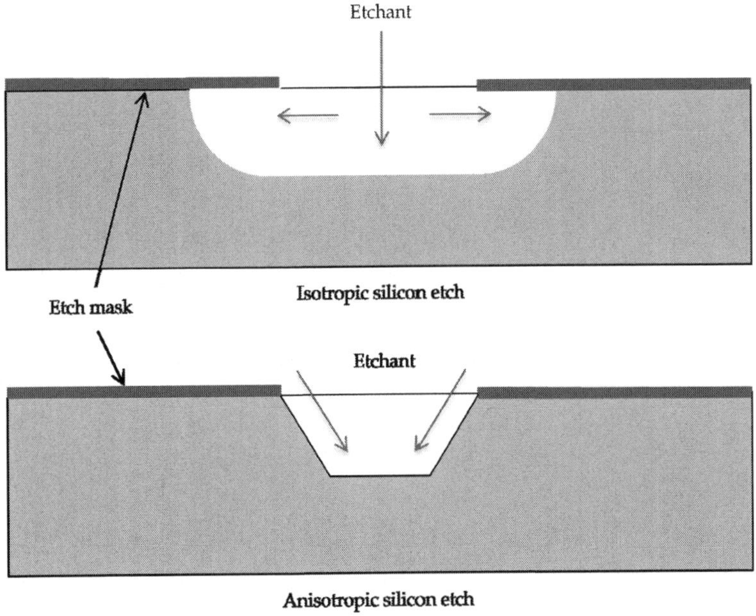

Figure 3.20 Wet etching of silicon.

Table 3.7 Wet-etch simulator

Wet etch			Etching rates	
			Material	nm/minute
Material	SiO_2 PECVD			
Rate	200	nm/minute	Thermal SiO_2	50
Thickness	1	micron	PECVD SiO_2	200
Time	5	minutes	Si_3N_4	100
Load/unload time	1	minute	Aluminium	200
Total time	6	minutes		
Batch/hour	10			
Wafers/batch	25			
Throughput	250	wafers/hour		

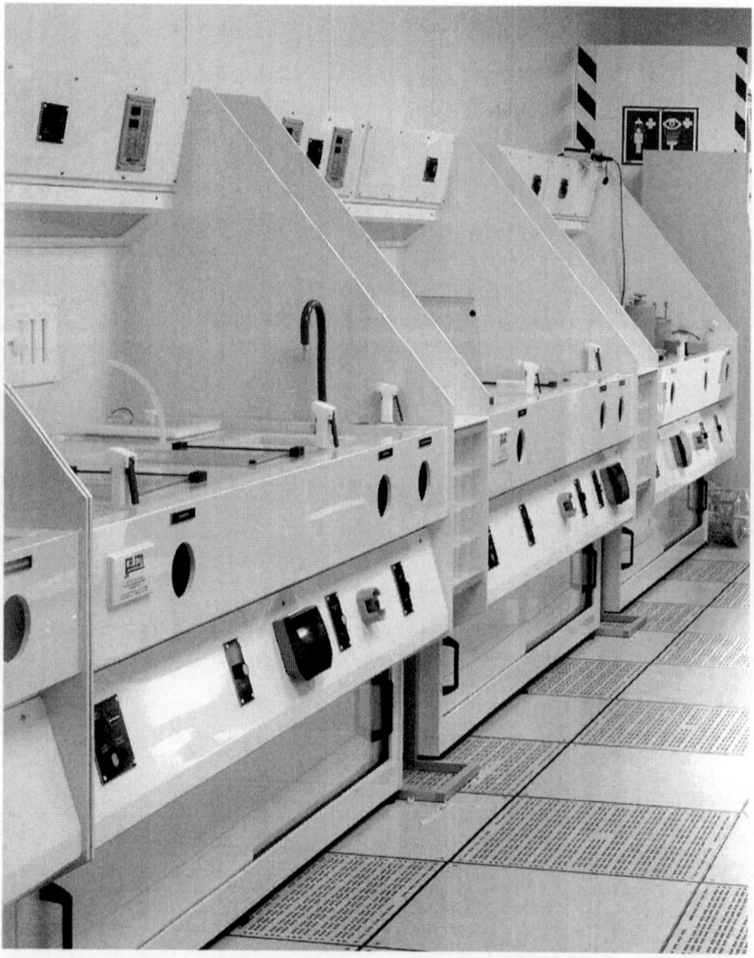

Figure 3.21 Wet-etch equipment (courtesy of the Southampton Nanofabrication Centre).

Isotropic etching etches equally in all directions, forming a trench that is wider than the defining mask and with curved sides. Anisotropic etching effectively maintains the top width defined by the mask and produces sloping sides of a fixed angle. Anisotropic wet-chemical etching is the basis of the high-aspect-ratio (HAR) microfabrication technique known as bulk micromachining (Elwenspock et al., 1998) and is discussed further in Chapter 4.

A typical wet-etch installation is shown in Fig. 3.21. The wet-etch simulator used in MEMSCOST to calculate throughput is shown in Table 3.7 for typical materials used in MEMS fabrication. The etch rates shown are for a buffered hydrofluoric acid (HF) etchant, a typical solution of 6:1 ratio by volume of 49% HF and ammonium fluoride (NH_4F) in water. The rates shown are averages from the literature (Williams, 2003; Witrouw, 2000). The rates depend on the solution ratio and can be increased by raising the temperature.

3.9 Reactive-Ion Etching

Reactive-ion etching (RIE) offers some significant advantages over wet-chemical etching, including an ability to etch isotropically or anisotropically at higher resolution on a range of materials without

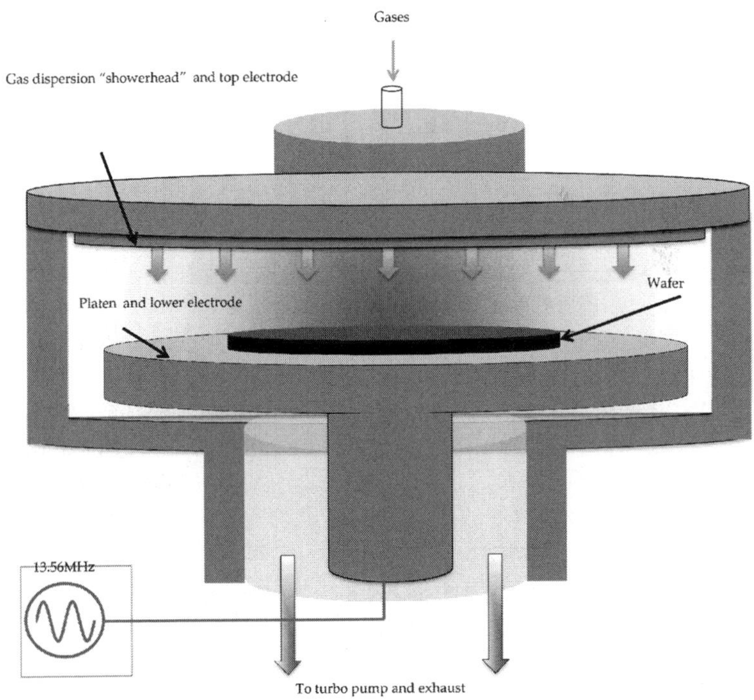

Figure 3.22 Schematic of a reactive-ion etcher.

Table 3.8 RIE simulator

Reactive-ion etch			Etching rates Material	nm/minute	
Material	SiO_2		Silicon (anisotropic)	300	Cl_2, BCL_3, H_2
Rate	25	nm/minute	Silicon (isotropic)	110	SF_6/O_2
Thickness	1	micron	Polysilicon	150	Cl_2, BCL_3, H_2
Time	40	minutes	SiO_2	25	CHF_3/O_2
Load/unload time	1	minute	Si_3N_4	675	CHF_3/O_2
Total time	41	minutes	Photoresist	100	O_2
Hours/batch	1.46		Aluminium	150	BCL_3, Cl_2, CH_4
Wafers/batch	1		Tungsten	100	CH_4, SF_6
Throughput	1.46	wafers/hour			

Abbreviations: BCL_3, boron trichloride; SF_6, sulphur hexachloride; O_2, oxygen; CHF_3, trifluormethane; CH_4, tetrafluormethane.

restriction to crystallographic planes. However, compared with wet-etching equipment, RIE is more expensive.

The basic construction of an RIE machine is shown schematically in Fig. 3.22. Plasma is created in the reaction chamber by an RF generator, and gas is introduced. The gas molecules are ionised, and ions and radicals are accelerated towards the platen, on which is placed the substrate to be etched. The ions cause a chemical reaction to occur at the substrate surface, although there is some physical (sputtering) removal of material. As the ions tend to strike the substrate vertically, RIE can produce etched troughs with reasonably anisotropic wall angles. A typical RIE machine is shown in Fig. 3.23. The RIE simulator used in MEMSCOST to calculate throughput for a given etchant is shown in Table 3.8 for the typical materials used in MEMS fabrication.

An RIE machine will undertake most tasks where thin films are involved and the etch rates are acceptable. Some limitations of RIE are set by the low pressure (e.g., 100 mTorr) and low power (100 W) at which the system operates (Legtenberg et al., 1995; Jansen et al., 1996).

Figure 3.23 RIE machine—Oxford Instruments Plasmalab 80+ (courtesy Imperial College).

An alternative configuration, where a higher aspect ratio and faster etching is required, is inductively controlled plasma (ICP) etching, which operates in a high-plasma-density mode, typically at a power of 1,500 W and a pressure of 10 mTorr. Gas flow is introduced at the top of the machine and is converted into plasma as it passes through the ICP RF source. The substrate to be etched is seated on a platen, which can be separately biased to improve the etching characteristics (McAuley et al., 2001). A schematic of an ICP machine is shown in Fig. 3.24 and a typical machine in Fig. 3.25. The ICP simulator used in MEMSCOST to calculate throughput is shown in Table 3.9 for typical materials used in MEMS fabrication.

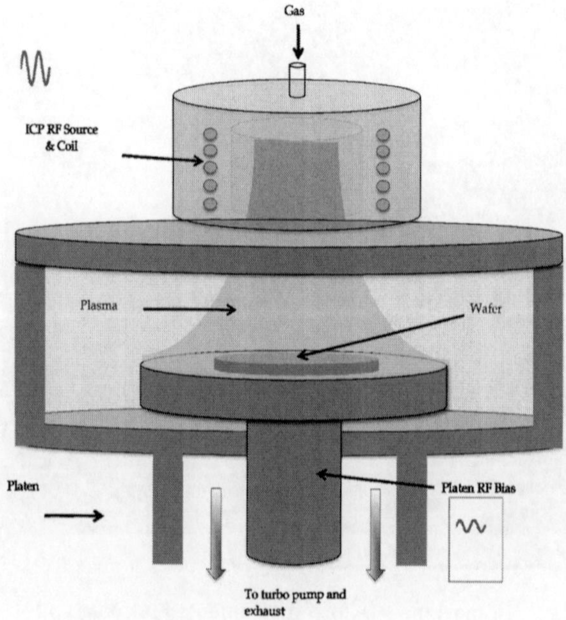

Figure 3.24 Schematic of an ICP etch machine.

Figure 3.25 Oxford Instruments Plasmalab 100 (courtesy of the Southampton Nanofabrication Centre).

Table 3.9 ICP etcher

ICP etch			Etching rates Material	nm/minute	Gases
Material	Polysilicon		Silicon	1,400	SF_6
Rate	2200	nm/minute	Polysilicon	2,200	SF_6
Thickness	0.5	microns	SiO_2	280	CHF_3
Time	0.23	minutes	Si_3N_4	190	SF_6, CF_4
Load/unload time	2	minutes	Aluminium	172	$HBr:BCl_3$
Total time	2.23	minutes			
Hours/batch	0.04				
Wafers/batch	1				
Time /wafer	0.04				
Throughput	25.0	wafers/hour			

3.10 Chemical Mechanical Polishing

Chemical mechanical polishing (CMP) is used to planarise films where the deposition process has left a sufficient out-of-plane area to affect, potentially, the quality and yield of subsequent processing. IBM developed CMP for semiconductor fabrication in the 1980s, but it was soon realised that many of the problems being solved were arising also in MEMS manufacture. As tighter and tighter tolerances were required for MEMS moving parts, the need to planarise became important, for example, to improve the depth of focus (DoF) for optical aligners.

The basic action of CMP is to mechanically polish the rotating wafer to be planarised using an orbital motion. CMP uses a polishing pad covered with chemical slurry that is dispensed continuously. The basic components are shown in Fig. 3.26 and a commercial machine in Fig. 3.27.

Material removal is caused, firstly, by the chemicals in the slurry loosening or reacting with the material and, secondly, by abrasives in the slurry. Finally, the polishing action removes the loosened material from the wafer (Luo, 2001; Woodie, 2007).

The rate of polishing depends on the downward pressure between the wafer and the polishing pad, the speed of rotation of the wafer and the platen and the action of the slurry chemicals.

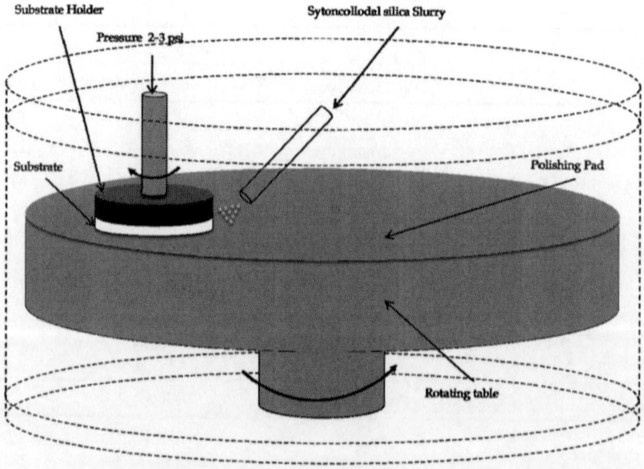

Figure 3.26 Schematic of chemical mechanical polishing.

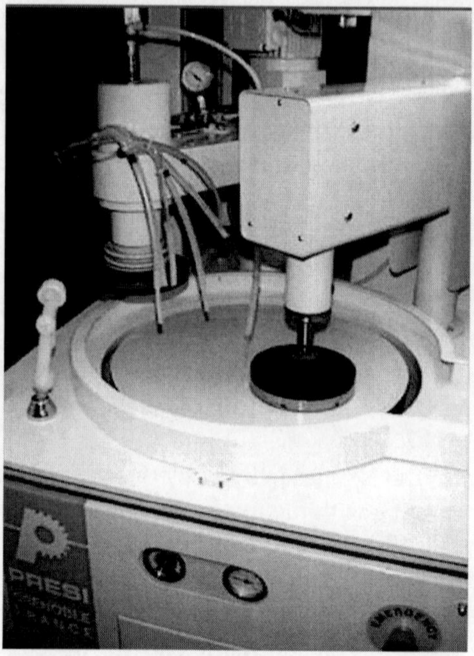

Figure 3.27 CMP machine (courtesy SMC, Edinburgh).

Table 3.10 CMP simulator

CMP			Polishing rates	
			Material	nm/minute
Polysilicon			Silicon	1
Rate	400	nm/minute	Polysilicon	400
Thickness	1	μm	SiO_2	200
Time	2.5	minutes	Si_3N_4	100
Load/unload time	1	minute	Photoresist	100
Total time	3.5	minutes	Aluminium	450
Batch/hour	17.1		Copper	150
Wafers/batch	1		Tungsten	200
Throughput	17.1	wafers/hour		

Pressures vary from 1 to 8 psi, with a polishing pad rotation velocity from 30 to 200 rpm (Hocheng et al., 2001). The polishing rate (R_{CMP}) is given approximately by Eq. 3.23 (Steigerwald, 1997) and is the basis of the CMP simulator shown in Table 3.10.

$$R_{CMP} = (kP + \alpha) + \beta \tag{3.23}$$

where:

P = pressure on the wafer (psi)
V = rotation speed of the polishing pad (rpm)
α = constant of proportionality
β = chemical removal rate
k = Preston coefficient of proportionality

3.11 Electron-Beam Pattern Generation

An electron-beam lithography (EBL) machine is a very accurate, high-resolution sub-micron pattern generator for commercial mask making and nanolithography applications. The capability of an EBL machine depends on the electron optical beam column design, the energy of the electron beam (50–100 KeV) and the scanning strategies used to expose large areas. In all cases the electron beam can only cover a relatively small field, limited to a few millimeters, before intrinsic aberrations distort the x–y position of the beam. To cover typical substrates (e.g., 125–175 mm mask/reticle plates

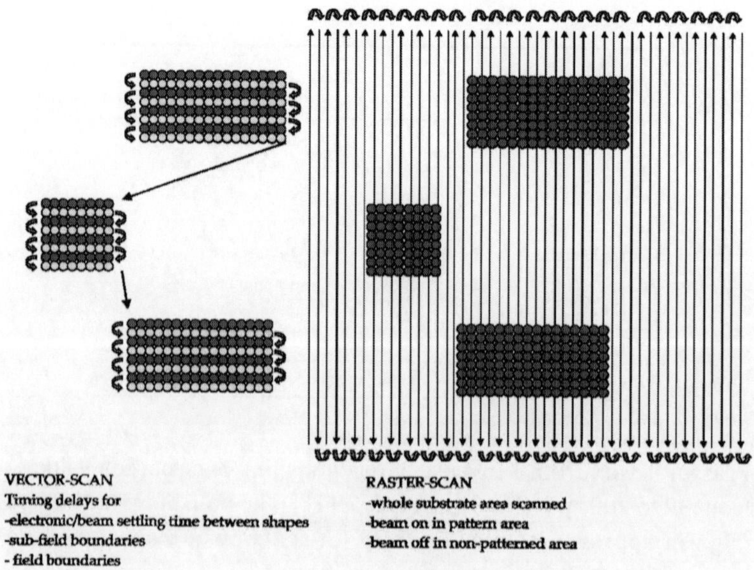

Figure 3.28 Scanning strategies for Gaussian beam EBL machines.

or 200–300 mm wafer diameters), the substrate is moved under the electron beam by a laser-digitised mechanical stage. This movement may be step-and-repeat (similar to an optical stepper) or continuous, leading to write-on-the fly strategies. To align the position of the axis of the electron beam to the digitised stage, a feedback signal, based on the stage position, is connected to deflection coils on the column.

Figure 3.28 shows the scanning strategies of two of the machine types that have been available over the last 30 years. The 'vectorscan' machine focuses an electron beam to a few nanometers (with a Gaussian beam profile) and deflects the beam to the starting position of a shape to be exposed and then scans the beam at high speed in the *x*- and *y*-directions. The beam is then blanked, while the beam position is vectored to the starting position of the next shape. The beam energy is 10–100 KeV and is focused down to a diameter of 5–1,000 nm. The beam is steered over a small, square field of 0.1–5 mm by electromagnetic coils at beam-stepping rates up to 50 Mhz, and the beam is switched on and off, as appropriate, to form

Table 3.11 EBL machine simulator

Parameter		Dimension
Wafer diameter	200	mm
Die size X	2	mm
Die size Y	2	mm
Field size (F)	1.024	mm
Pixel size (p)	0.1	microns
No. of pixels	1.05E8	
Fractional field coverage (ρ) %	10,500,000	10
Current (I)	18.4	nA
Resist sensitivity (S)	5	uC/cm^2
Maximum scan frequency	50	MHz
Scan frequency	36.8	MHz
Exposure time per field	0.285	seconds
Field-to-field time (T_{FF}) (inc. registration)	0.032	seconds
Time/field	0.317	seconds
No. of fields/die	4	
Time/die	1.268	seconds
No. of dies/wafer	6,656	
Load/unload (T_U)	0	seconds
Total exposure time	8,439.8	seconds
Substrates/hour	0.43	

images in a resist. The vectorscan EBL machine is the most accurate, high-resolution design but suffers from low throughput, typically less than a wafer per hour (see simulator example in Table 3.11).

A much simpler machine has a beam with a Gaussian beam profile and a raster scan that deflects the beam in one dimension only, avoiding complex two-dimensional electron optical deflection, while the substrate is moved continuously in one direction by a laser-digitised stage (not shown). While this type of machine has less resolution capability compared with a vectorscan machine, its sub-micron resolution is well suited to mask and reticle manufacture, and it has been the mainstream commercial mask maker for the last 30 years (Eynon & Wu, 2005).

The vectorscan strategy has been further developed by projecting an electron image of an aperture rather than using a focused beam (see Fig. 3.29). A variable shaped beam, typically 1.5 mm × 0.5 mm, can be projected onto a substrate in a single exposure and

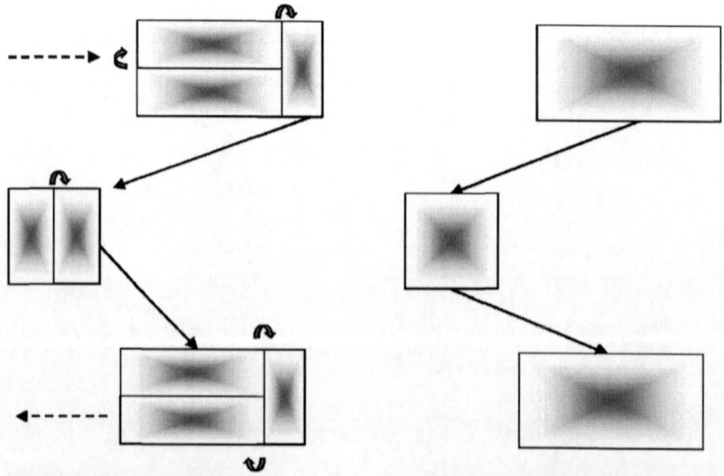

Figure 3.29 Scanning strategies for variable-shaped beam and cell projection EBL machines.

repeated several times to produce a complex shape. Thus a few variable-shaped beams will cover the same area that would require multiple focused Gaussian beams and will reduce the overall write time per substrate significantly (Hahmann et al., 2007). The beam is then vectored to the next position (Stolberg, 2007). A more advanced development to improve throughput even further has been the concept of cell projection, where an image of one of a number of standard-shaped apertures is projected onto the wafer (Fig. 3.29).

The time to expose a layer of resist will depend on many parameters. Firstly, the electron beam must provide sufficient charge in the resist to expose an image, that is, the product of beam current and exposure time must be greater than the resist sensitivity ($\mu C/cm^2$). Secondly, the beam will need time to fill in the designed feature and to step from feature to feature. Thirdly, the stage will need time to traverse from field to field. Finally, time will be necessary to load/unload the wafer. As an example of throughput, consider the vectorscan Gaussian beam machine.

Time/field = Number of beam exposures/Field × % of field written × Time per beam exposure + Time to move the beam from one field to another

$$\text{Time/field} = F^2\rho/(p^2 M) + T_{\text{FF}} \tag{3.24}$$

$$\text{Time/die} = A\rho/(p^2 M) + T_{\text{FF}} A/F^2 \tag{3.25}$$

$$\text{Time/wafer} = N_{\text{D}}[A\rho/(p^2 M) + T_{\text{FF}} A/F^2 + T_{\text{DD}}] \tag{3.26}$$

where:

F = field size (mm)
ρ = fraction of mask/wafer actually addressed by the electron beam
p = pixel size (μm)
M = beam-stepping frequency (MHz)
T_{FF} = time to move from field to field (seconds)
A = die area (mm^2)
N_{D} = number of die per wafer
T_{DD} = time to move from die to die, including die registration (seconds)

(NB: With a multi-wafer cassette, the load/unload time may be ignored.)

The current available in the beam, for the pixel size chosen, is limited by beam aberrations (Brewer, 1980). The beam-stepping frequency is then determined by the sensitivity of the electron-beam resist up to a maximum set by the beam deflection electronics.

$$M < (F/p)^2 I\rho/S \tag{3.27}$$

where:

I = electron-beam current (nA)
S = resist sensitivity ($\mu\text{C/cm}^2$)

A schematic of a basic vectorscan EBL machine is shown in Fig. 3.30 and consists of the electron-beam column and the laser digitised stage previously discussed. The electron-beam column consists of an electron source (the gun), a series of focusing coils, a beam deflector and a blanking unit to turn the beam off according to

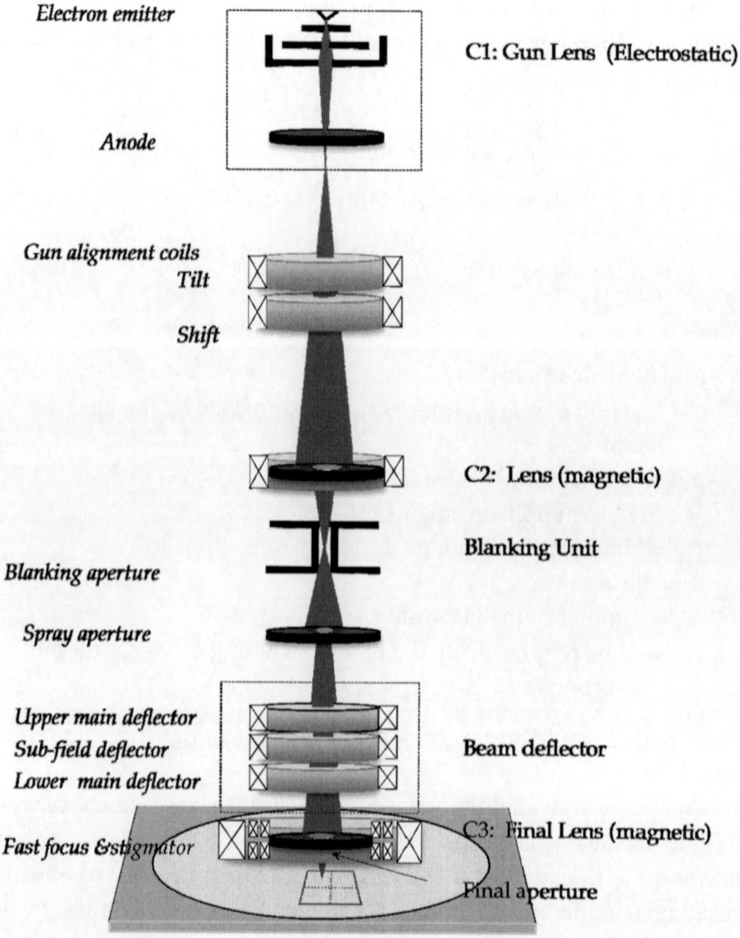

Figure 3.30 A Gaussian beam vectorscan electron-beam column.

the computer instructions. The size of the beam is formed by a series of apertures at different positions on the column. The final aperture defines the semiangle of the optics and hence the beam current–diameter relationship. A commercial Gaussian beam, vectorscan machine is shown in Fig. 3.31.

Table 3.11 shows the MEMSCOST EBL Simulator set up for a 200 mm wafer and 2 mm × 2 mm die. These settings are mainly to provide the sub-micron resolution required for ICs.

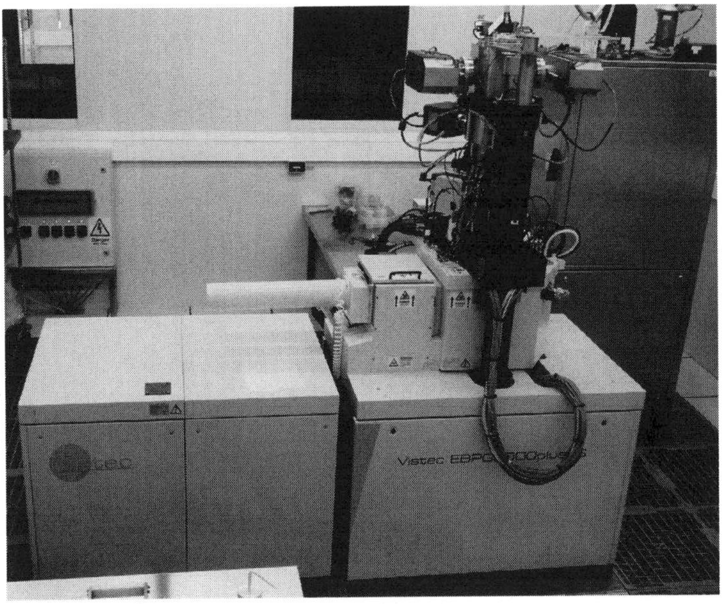

Figure 3.31 EBL machine—Vistec EPG5000plus (courtesy Monash University).

However, MEMS fabrication requires much lower resolution so that by increasing the beam diameter and the field size, significant throughput improvements can be obtained.

At the time of writing, electron-beam machines are rarely used for MEMS devices, as the machines are expensive, the throughput is too low and extreme accuracy and resolution are not (yet?) required for production manufacture. EBL may find a role when MEMS becomes nanoelectromechanical systems (NEMS) or when significant research and development (R&D), and hence mutiple design changes, becomes necessary and the costs of mask changes become prohibitive.

3.12 Surface Microengineering Equipment Summary

Table 3.12 summarises the capital cost for a typical portfolio of surface microengineering machines assumed for MEMSCOST

Table 3.12 Capital costs and throughput rates for surface microengineering for a nominal 1 μm thin-film thickness

Surface microengineering equipment and processes	Capital per $/unit	Throughput Wafers/hour	Maximum
Litho aligner	250,000	40	302,400
Litho stepper (0.35 μm)	3,500,000	41.6	314,496
Litho stepper (0.18 μm)	6,000,000	41.6	314,496
Resist processes (thin film)	800,000	357	2,698,920
LPCVD poly (per stack)	300,000	14.9	112,644
LPCVD silicon nitride	300,000	13.4	56,700
PECVD SiO_2	1,000,000	10	75,600
Thermal oxide/stack	300,000	11.6	87,696
DC sputter (aluminium)	1,500,00	1.13	8,543
Wet etch (oxide)	1,000,000	250	1,890,000
Dry etch (RIE)	1,000,000	1.46	11,038
Dry etch (ICP)	1,500,000	25	189,000
HF release	250,000	250	1,890,000
CMP	750,000	17.1	129,276
Electron beam	3,500,000	0.43	3,251

calculations. To compare the throughput of each machine the machine simulators assume a film thickness of 1 μm. However, when calculating costs for a device, the film thickness specified in the process list is used (see Chapter 8). Also shown in Table 3.12 are the maximum process steps each piece of equipment can perform in a year with 90% uptime. In most cases one of each machine is sufficient to support the annual wafer requirements for a given MEMS device.

References

ASML (2012). PA5500 series steppers. www.asml/show.do?ctx=6644.

Bhatt V, and Chandra S (2007). Silicon dioxide films by RF sputtering for microstructure fabrication in MEMs. *J. Micromech. Microeng.*, **17**, 1066–1077.

Bhatt V, and Chandra S (2009). Silicon nitride films deposited by RF sputtering for microelectronics and MEMs applications. *J. Electron. Mater.*, **38**(9), 1979–1989.

Brewer, GR (1980). *Electron Beam Technology in Microelectronic Fabrication.* Academic Press.

Brigham Young University (2012). Oxide growth thickness calculator. www.cleanroom.byu.edu/OxideThickCalc.phtml

Chandra S, Bhatt V, and Singh R (2009). RF sputtering: a viable tool for MEMs fabrication. *Sadhana,* **38**(4), 543–556.

Canon (2012). I-line and deep UV steppers. usa.canon.com/cusa/semiconductor/products/semiconductor_equipment

Deal RE, and Grove AS (1965). General relationship for the thermal oxidation of silicon. *J. Appl. Phys.*, **36**(12), 3770–3778.

Eynon BG, and Wu B (2005). *Photomask Fabrication Technology.* McGraw Enginneering Series.

Elwenspoek M, and Jansen H (1998). *Silicon Micromachining (Cambridge Studies in Semiconductor Physics and Microelectronic Engineering).* Cambridge University Press. ISBN: 0-521-59054-X(hb), 11–31.

Gerloch, G, et al. (2008). *Introduction to Microsystem Technology.* John Wiley & Sons. ISBN: 0470770929, 81–82.

Grove AS (1971). *Physics and Technology of Semiconductor Devices.* John Wiley & Sons. ISBN: 0471329983.

Hahmann P, et al. (2007). High resolution variable-shaped beam direct write. *Microelectron. Eng.*, **84**, 774–778.

Hocheng H, et al. (2001). Modeling and experimental analysis of the material removal rate in the chemical mechanical planarization of dielectric films and bare silicon wafers. *J. Electrochem. Soc.*, **148**(10), G581–G586.

Hutcheson, DG (2005). Moore's law: the history and economics of an observation that changed the world. *Electrochem. Soc. Interface,* **14**(1) (Spring 2005), 17–21.

Jaeger R (2002). *Film Deposition: Introduction to Microelectronic Fabrication.* Prentice Hall. ISBN: 0-201-44494-7.

Jansen H, et al. (1996). A survey of the reactive ion etching of silicon in microtechnology. *J. Micromech. Microeng.*, **6**, 14–28.

Junghoon J, et al. (2000). Low-temperature polysilicon deposition by ionized magnetron sputtering, *J. Vac. Sci.Technol. A,* **18**(4), 2006–2011.

Legtenberg et al. (1995). Anisotropic reactive ion etching of silicon using $SF_6/O_2/CH_3$ gas mixtures. *J. Micromech. Microeng.*, **6**, 14–28.

Levinson, HJ (2011). *Principles of Lithography, Third Edition. SPIE Monograph,* Vol. PM198.

Linder C, et al. (1992). Surface micromachining. *J. Micromech. Microeng.*, **2**(3), 122–132.

Luo J, and Dornfeld D (2001). Material removal mechanism in chemical mechanical polishing: theory and modeling. *IEE Trans. Semicond. Manuf.*, **14**(2).

Nikon (2012). NSR series steppers. www.nikon.com/products/precision

Madou, MJ (2011). *Fundamentals of Microfabrication and Nanotechnology, Third Edition*. CRC Press. ISBN-10: 0849331803, ISBN-13: 978-0849331800.

McAuley SA, et al. (2001). Silicon micromachining using a high density plasma source. *J. Phys. D: Appl. Phys.*, **34**, 2769–2774.

Microchemical (2010). Exposure of photoresists. www.microchemicals.eu/technical_information.

Rai-Choudhury P (1997). Microlithography, micromachining and microfabrication. *IEE Mater. Devices*, **12**(1), microlithography.

Steigerwald JM, et al. (1997). *Chemical Mechanical Planarization of Microelectronic Materials*. John Wiley & Sons.

Stolberg, I (2007). Visitec shaped beam technology. www.fasimit.de.fileadmin/…/Grundlagen_EBeam_technology.pdf

SUSS (2011). The lithography benchmark from R&D to 3D MEMS production. www.suss.com/mask-aligner.

Sze, S (2001). *Semiconductor Devices: Physics and Technology*. 2nd ed. John Wiley & Sons. ISBN: 0-471-062753.

Woodie, D (2007). *Chemical Mechanical Polishing (CMP)*. Cornell Nanoscale Facility, Cornell University.

Williams, K, et al. (2003). Etch rates for micromachining processing – part II. *J. MEMS*, **12**(6).

Witvrouw, A, et al. (2000). B A comparison between wet HF etching and vapor HF etching for sacrificial oxide removal. *Proc. SPIE*, **4174**, 130–141.

Chapter 4

High-Aspect-Ratio Micromachining

4.1 Introduction

This chapter will detail the equipment and processes that have been specifically developed for microelectromechanical systems (MEMS) fabrication and address the issue of the relative magnitude of the forces acting on the microstructures. MEMS devices differ from semiconductors in that the microstructures require attention not only to the planar or x–y dimensions but also to the z-dimension or depth. Many materials must be selected for their mechanical rather than electrical properties.

Generally the minimum x–y dimensions for MEMS is in the region of 1–5 μm and often much larger (50–100 μm), compared with minimum dimensions of 10–100 nm for a semiconductor device. Similarly, the z-dimension of a MEMS device will be much larger than for a semiconductor device, 5–500 μm, compared with 10–100 nm. The special techniques developed for MEMS fabrication are referred to as 'high aspect ratio' (HAR) and have aspect ratios up to 1:500, compared with a typical ratio of 1:10 for semiconductor devices.

The fabrication of HAR microstructures may require attention to the sidewall angle and its roughness. For example, a deep structure

MEMS Cost Analysis: From Laboratory to Industry
Ron Lawes
Copyright © 2014 Pan Stanford Publishing Pte. Ltd.
ISBN 978-981-4411-06-6 (Hardcover), 978-981-4411-07-3 (eBook)
www.panstanford.com

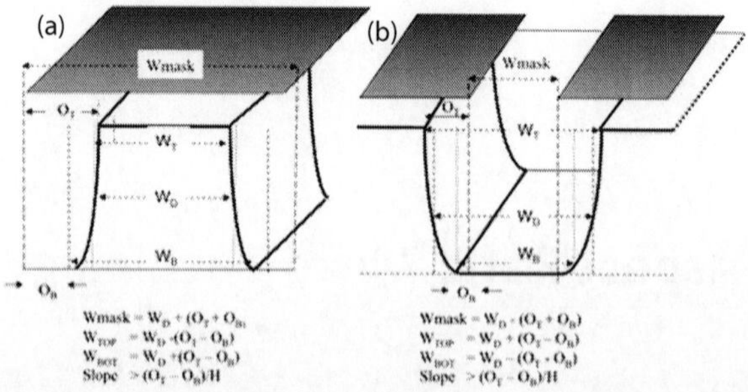

Figure 4.1 Schematic of deep microstructures formed by a lithography process as a negative (a) and a positive (b) version.

may require near-vertical wall angles if the top and bottom of the microstructure are to meet tight tolerances. It is therefore common in a MEMS design, for example, a comb actuator or sensor, to consider the three-dimensional tolerances of the microstructure. Figure 4.1a illustrates the negative form and Fig. 4.1b the positive form of an HAR resist structure and shows the dimensions that affect the quality of subsequent processes.

The performance of a MEMS device may depend upon the absolute and relative values of various forces, which will vary as a function of decreasing dimension. Consider the space shown in Fig. 4.2 on which various forces may act, where B, L and s are the linear dimensions in the x-, y- and z-directions, respectively.

If the volume outlined were filled with a solid mass of material (Fig. 4.2a) then the gravitational force (F_G) would be

$$F_G = 9.8\rho LBs \tag{4.1}$$

where:

ρ = density of material (kg/m^3)
g = acceleration of gravity (9.81 m/second^2)

If two electrodes, each of area $L \times B$ were separated in air by a distance s and a voltage V applied (Fig. 4.2b), then the force between the electrodes would be

$$F_E = \varepsilon_O LBV^2/(2s^2) \tag{4.2}$$

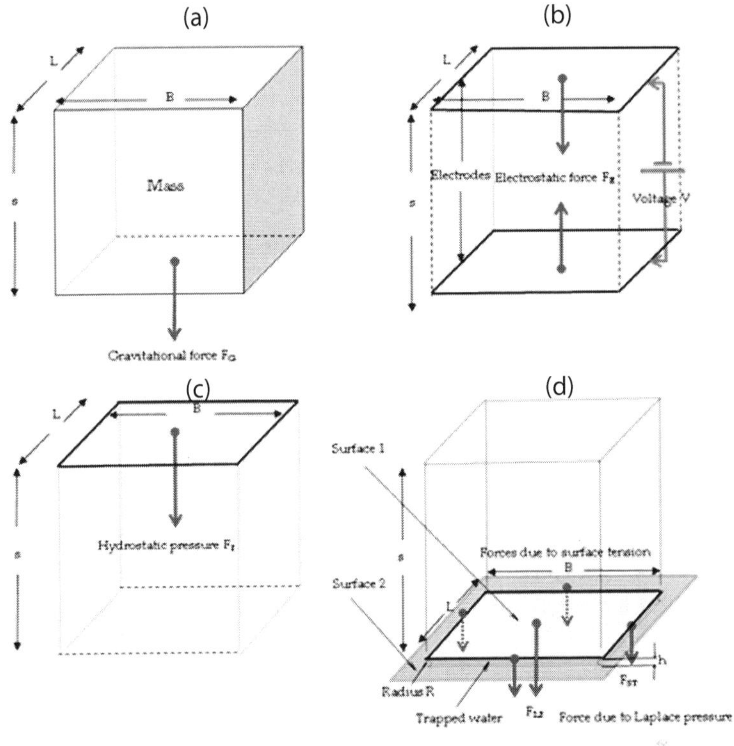

Figure 4.2 Notional space in which various microforces can act.

where:

ε_0 = permittivity of a vacuum = 8.85×10^{-12} Fm^{-1}
V = the voltage across the plates

Note that a typical comb actuator or sensor would have n electrodes, thereby increasing the total force/voltage by n.

If the same area were subjected to a pressure of p kg/m^2 (Fig. 4.2c) then the force generated (F_P) would be

$$F_P = \rho LB \tag{4.3}$$

If the solid volume LBs had an area $A = LB$ in contact with a liquid (Fig. 4.2d), then the forces acting on the area are due to the Laplace pressure (F_{LP}) and a surface tension force (F_{ST}) (Syms et al., 2003). In general the Laplace force dominates. Ignoring the depth of immersion in the liquid the total force (F_T) on this area would be

$$F_T = \iint P_{LP} dA + \int F_{ST} dl \quad (4.4)$$

where:

γ = surface tension coefficient
P_{LP} = Laplace pressure
$\int dl$ = periphery of the surface in contact with a liquid
$\int dA$ = area of the surface in contact with a liquid

$$F_T = \gamma LB/R + 2\gamma(L + B)\cos\theta \quad (4.5)$$

where:

R = radius of curvature of the liquid meniscus
θ = liquid contact angle at edge of surface

The Laplace pressure shown in Eq. 4.5 only applies to the centre of the surface edge (where there is one radius) and not to the corners, where there are two radii. In the case of the surface tension component, the value of θ is unknown. Accurate determination of F_T would require numerical analysis, which is not pursued here. To calculate an order of magnitude for F_T Eq. 4.5 is used, as shown with $\cos\theta = 1$. For the example shown in Fig. 4.2d, assuming the liquid is trapped between a microstructure (surface 1) of area $L \times B$ and another structure (surface 2), then the radius R is approximately $h/2$, where h is the separation of the two surfaces. In a MEMS structure, this separation may be only a few microns and often be defined by a sacrificial layer dimension.

An indication of the relative values of some of the forces involved in microengineering can be seen in Fig. 4.2 and then scaling conformally, that is, keeping the ratio between L, B and s constant

$L = 100\ \mu m$
$B = 20\ \mu m$
$s = 5\ \mu m$
$\sigma = 2{,}330\ kg/m^3$ for silicon
$\gamma = 0.072\ N/m$ for water
$p = 2 \times 10^5\ N/m^2 = 29$ psi

(NB: pressure chosen to demonstrate the magnitude of the hydrostatic force)

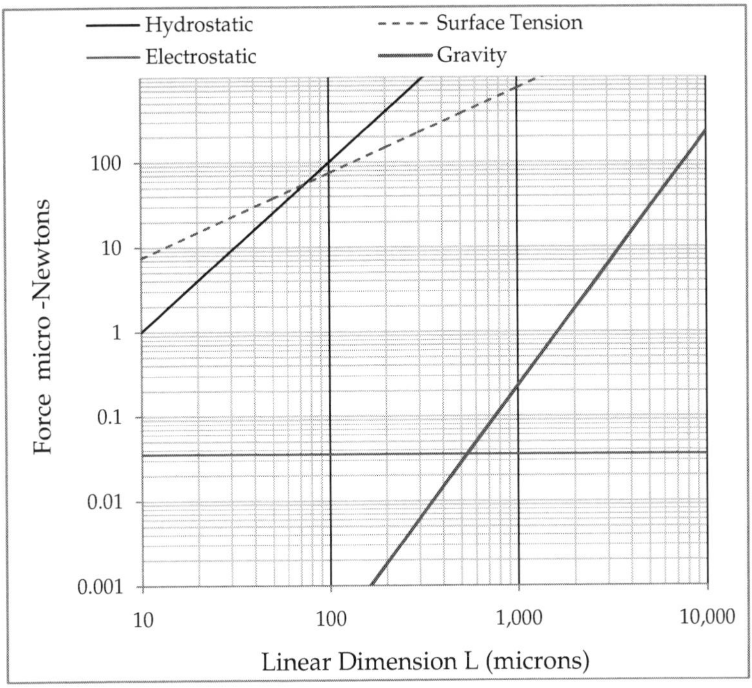

Figure 4.3 Non-conformal scaling of forces with dimension.

Figure 4.3 illustrates that the gravitational and electrostatic forces are small compared with those generated by the presence of a liquid. The relatively high value of the force due to the Laplace pressure leads to problems with MEMS manufacture, as stiction between any trapped water after cleaning and a mechanical structure, such as a comb sensor, will cause malfunctioning of such delicate microstructures. This illustrates why there are problems with stiction when wet-cleaning micro-parts and hence the need for critical-point-drying (see Section 4.9 later in this chapter).

4.2 Deep Reactive-Ion Etching

The extension of the basic reactive-ion etching process is known as deep reactive-ion etching, or DRIE, and is widely used in the MEMS industry to etch high-quality thin films. The practical version is known as the Bosch process after the company inventors (Laermer & Schilp, 1994; Laermer & Urban, 2005). DRIE is a form of plasma etching that produces sub-micron microstructures with depths up to 500 μm and with near-vertical sidewalls. It uses conventional lithography steps to define features in a suitably robust masking layer (e.g., resists, metal) on the surface of the wafer.

After each etch step the sidewall is coated with a chemically inert material to prevent erosion of the sidewall (a process known as passivation) during the next etch step. The reactive etchant is fluorine radical, typically derived from a mixture of sulphur hexafluoride (SF_6) and oxygen gas (O_2). The passivation material is derived from octofluorocyclobutane (C_4F_8).

The etch-passivation sequence is repeated until the microstructure reaches the required depth. The passivated sidewall is immune to chemical action and is not subject to the directional ion bombardment. The base of the etched feature is subjected to both chemical and directional ion bombardment, hence the near-vertical sidewall profile. The DRIE process leaves minute 'scallops' on the wall of the microstructure (see Fig. 4.4) whose size depends on the vertical etch distance per cycle. Therefore the time to fabricate a given depth will not only depend on machine parameters, such as the chamber pressure, but also depend on the required surface quality of microstructure walls, that is, the smaller the scalloped dimensions, the longer the etch time to a given depth. There will also be an undercutting of the resist at the mask interface, which will increase the nominal dimension at the surface of the etched microstructure.

DRIE is now an essential process for the MEMS industry due to its ability to etch HAR microstructures in a variety of materials with near-90° wall angles. DRIE is a key process for fabricating capacitive sensors that require a deep structure with only a few microns between the electrodes. The example in Fig. 4.5 shows a DRIE-etched comb sensor/actuator in silicon with a gap of 4 μm between 100 μm deep electrodes (an aspect ratio of 25). Note that an optical wafer stepper can define sub-micron features for MEMS and would require DRIE rather than wet-chemical etching.

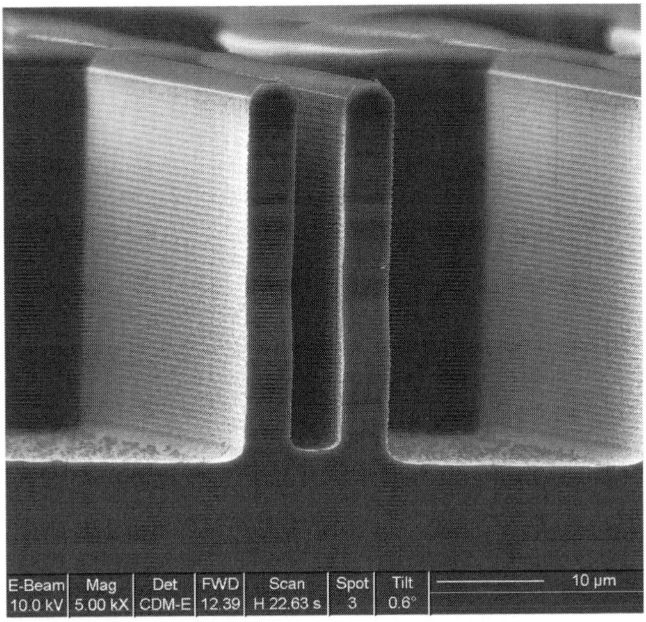

Figure 4.4 Sidewall roughness due to scalloping (courtesy School of Mechanical Engineering, Birmingham University).

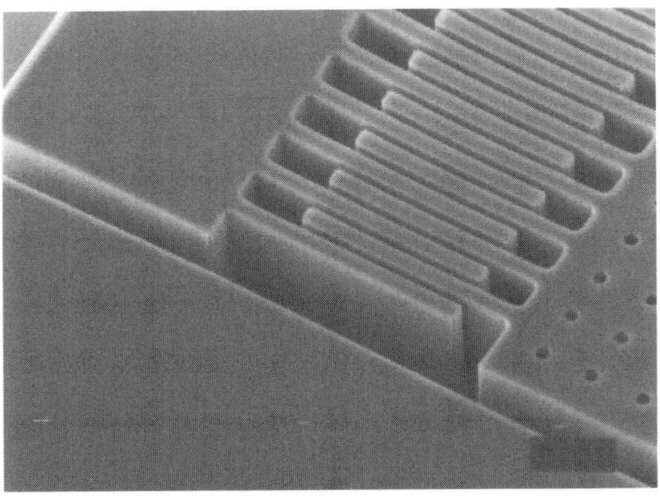

Figure 4.5 Silicon comb actuator/sensor fabricated by DRIE, SF6 etch with a 4 μm gap and a depth of 100 μm (courtesy Imperial College London).

Table 4.1 Typical DRIE silicon etch process

Parameters	Value	Units
Etch		
SF_6 etchant gas	130	sccm
O_2 gas additive	13	sccm
Chamber pressure during etching	35	mTorr
Etch rate (silicon)	3.4	μm/minute
Etch time/cycle	10	seconds
Passivation		
C_4F_8 passivation gas	80	sccm
Chamber pressure during passivation	18	mTorr
Passivation time/cycle	7	seconds
Coil power	600	watts
Platen bias power	20	watts
Platen temperature	25	°C
Net etch rate (silicon)	2	μm/minute

The actual etch rate and the dimensional tolerance with microstructure depth are a complex function of the aspect ratio of the microstructure and percentage loading (area etched/total area) and the etch/passivation cycle time. A typical process, producing good, smooth surfaces at an acceptable etch rate, is given in Table 4.1. The internal components of a typical DRIE machine to control etching and sidewall passivation are shown schematically in Fig. 4.6 and a laboratory machine in Fig. 4.7. With the latest machines, the etch rate is greatly increased (up to 40–50 μm per minute), allowing much higher throughput rates, as shown in the DRIE simulator used in MEMSCOST (Table 4.2).

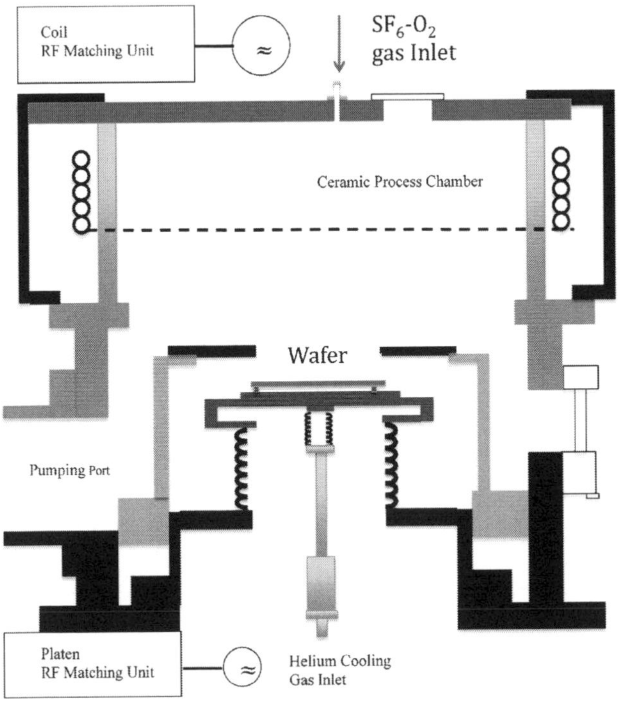

Figure 4.6 Schematic of a DRIE machine.

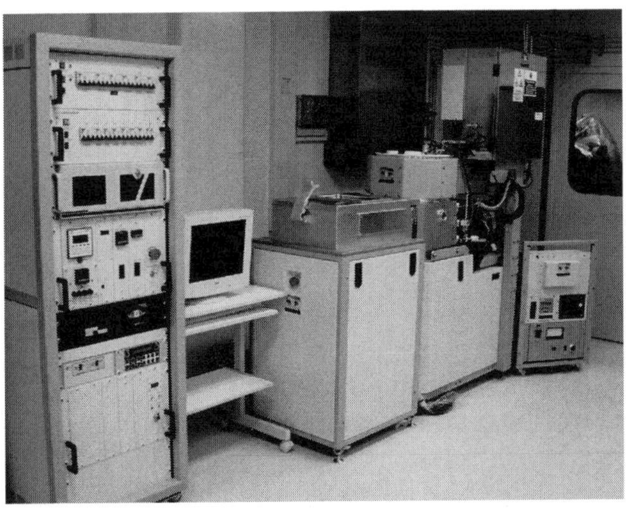

Figure 4.7 A DRIE machine: STS Single Chamber Multiplex (courtesy Imperial College).

Table 4.2 DRIE simulator for high-throughput etching

Silicon		
Etch depth	100	μm
Etch rate	20	μm/min
Etch time	10	seconds
Etch distance	3.3	μm
Lateral etch (estimated)	0.4	μm
Passivation time	5	seconds
Total time	15	seconds
Effective etch rate	13.2	μm/minute
No. of etch/passivate cycles	30	
Time	7.5	minutes
Load/unload	1	minute
Total time	8.5	minutes
Batches/hour	7.1	
Wafers/batch	1	
Throughput	7.1	wafers/hour

4.3 Excimer Laser Ablation

By the late 1970s, despite increases in the numerical aperture of lenses, the resolution of optical steppers with conventional mercury lamp sources was insufficient to provide the resolution required of the future technology nodes. The availability of the excimer laser at deep ultraviolet (UV) wavelengths was soon proposed as the light source to solve, in part, the resolution problem (Jain, 1982; 1990). The remainder of the solution was to use suitable deep UV lenses, which had to be developed by the industry. The excimer laser is currently used on all modern steppers (Lin, 2009).

It was soon realised that excimer lasers have sufficiently short pulses to cause photo-dissociation in polymers (Davis et al., 1993). This process is known as ablation and is a process that is significantly different to conventional lithography. Commercial exploitation of the excimer laser as a source for high-resolution micromachining soon followed (Rumsby & Gower, 1991) with a wide range of applications (Harvey et al., 1995). A schematic of the beam delivery system of an excimer laser machine is shown in Fig. 4.8 and a production machine installation in Fig. 4.9.

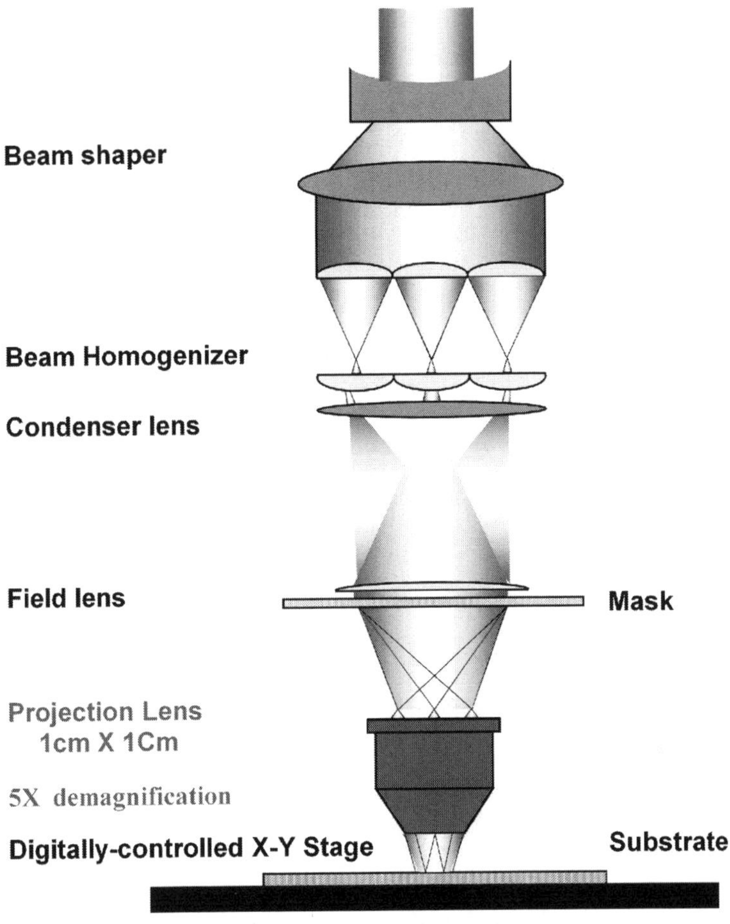

Figure 4.8 Schematic of an excimer laser ablation machine.

Excimer gas lasers can deliver ultraviolet radiation at wavelengths of 157–351 nm, depending on the gas mixture used. Krypton fluoride (KrF, λ = 248 nm) and argon fluoride (ArF, λ = 193 nm) produce the most useful wavelengths for micromachining, as the photon energies are close to the photo dissociation energies for polymers such as polyimide and polycarbonate.

The basic components of a laser ablation machine mimic those of an optical wafer projector in that there is a UV source (the excimer laser), a mask, a demagnifying lens (Goodall et al., 1990) and a

Figure 4.9 An Exitech excimer laser ablation machine (courtesy miniFAB).

stage system that carries one or more wafers. The resultant beam intensity contains inhomogeneities that could be transferred to the substrate. These inhomogeneities are much reduced by employing a beam homogeniser between the laser source and the mask.

The energy density of the laser source is increased by projecting an image of the illuminated mask through a demagnifying lens. Assuming a source energy density of between 200 mJ cm^2 and 300 mJ cm^2 and a typical energy loss of 40% in the beam delivery system, a demagnification of $\times 2$ will increase the beam energy density to between 480 mJ cm^2 and 720 mJ cm^2, which is sufficient to produce a useful ablation depth per pulse.

The depth of cut into a given material as a function of the energy density is shown in Fig. 4.10. Note that the depth per pulse is sufficiently accurate to enable the overall depth of ablation to be set simply by defining the number of pulses. As the laser beam is scanned (step-and-repeated) across the mask, the reduced image is scanned across the substrate. The write time will be set by the ratio of the substrate area to the beam area at the substrate. Throughput

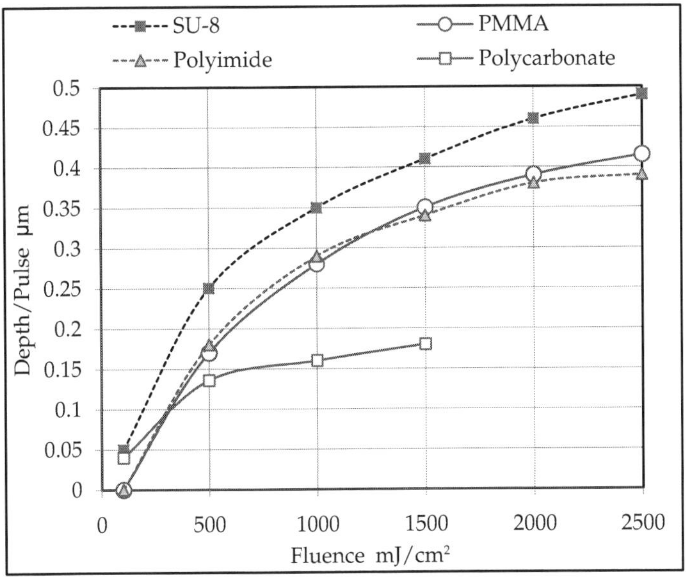

Figure 4.10 Depth per pulse as a function of the energy density at a 248 nm wavelength and 20 ns per pulse for a variety of materials.

can be increased using more powerful energy densities, although there is a limitation of about 500 mJ cm^2 set at the mask, if a standard chrome-on–glass mask is used.

MEMSCOST contains a laser ablation machine simulator (shown in Table 4.3) that takes into account the laser properties, the material to be micromachined and the time taken to machine to the required depth. The parameters shown are for the Coherent Lamda SX 300 K industrial standard laser source. In recent years, developments in ultrafast UV lasers, that is, nano, pico and femto pulse widths, compared with the excimer at 20 nS, have provided an alternative to the excimer laser, particularly for materials that are difficult to fabricate, such as silicon (Rizvi et al., 2001).

4.4 X-Ray LIGA

LIGA stands for Lithographie, Galvanoformung and Abformung (in German), meaning lithography, electroforming and moulding.

Table 4.3 Laser ablation machine simulator for a 248 nm wavelength

SU-8		
Required depth	100	μm
Wavelength	248	nm
Output of laser/pulse	1000	mJ
Pulse repetition rate	300	Hz
Area	4.76	cm^2
Energy loss through beam delivery	40	%
Energy density @ mask	126	mJ cm^2
Lens demagnification	2	
Energy density @ wafer	504	mJ cm^2
Depth cut/pulse	0.25	μm
Number pulses to depth	400	
Time per pulse	1.33	seconds
Area covered	1.19	cm^2
Wafer diameter	200	mm
Shots/wafer	105,600	
Time to cover wafer	352.0	seconds
Load/unload	60	seconds
Total time	412	seconds
Batches/hour	8.7	
Batch size	1	wafer
Throughput	8.7	wafers/hour

It is a lithography technique that uses X-rays from an electron synchrotron (Ehrfeld & Schmidt 1998). A typical X-ray synchrotron is the Diamond synchrotron, shown in Fig. 4.11, based in South Oxfordshire, UK. Synchrotrons are constructed and mainly used by national research institutes for basic research. Nevertheless several centres offer access to a synchrotron at a commercial rate to exploit the technology, notably Angstromquelle Karlsruhe (ANKA) (Saile, 2009), Berliner Elektronenspeicherring-Gesellschaft für Synchrotronstrahlung (BESSY) in Berlin (Loechal et al., 2007) and the Center for Advanced Microstructures and Devices (CAMD) in Baton Rouge.

Intermediate energy X-rays (3–10 KeV) will penetrate several centimetres into a resist material without significant scattering, leading to near-vertical wall angles and HAR microstructures. From

Figure 4.11 A synchrotron radiation facility (source: Diamond Light Source, South Oxfordshire, UK).

a cost point of view, their main disadvantage is the capital cost of a synchrotron ($200–$500 million) and the requirement for special gold-on-beryllium masks ($7,000–$10,000 per mask) with no commercial supplier (each research organisation fabricates its own masks).

LIGA operates by conventional resist exposure through a special mask, and the exposure time is dominated by the type and thickness of the absorbing resist layer. Initially, the LIGA resist was mainly polymethylmethacrylate (PMMA), which required 5–10 hours of exposure, but the advent of the SU-8 resist, which is 100 times more sensitive to X-rays than PMMA, has reduced the exposure time from hours to minutes. The actual exposure time is a complex calculation between the output current and spectrum of the synchrotron, the mask substrate and the resist type. This has been well-documented in simulation programs such as DoseSIM (Meyer et al., 2002; 2003), RALBEAM (Lawes, 2001), X3D (Hafizovic et al., 2003) and LEX-D (Tikare, 2005).

The X-ray beam is transmitted to the exposure station by a complex line of magnets (see Fig. 4.12) and arrives at the lithography

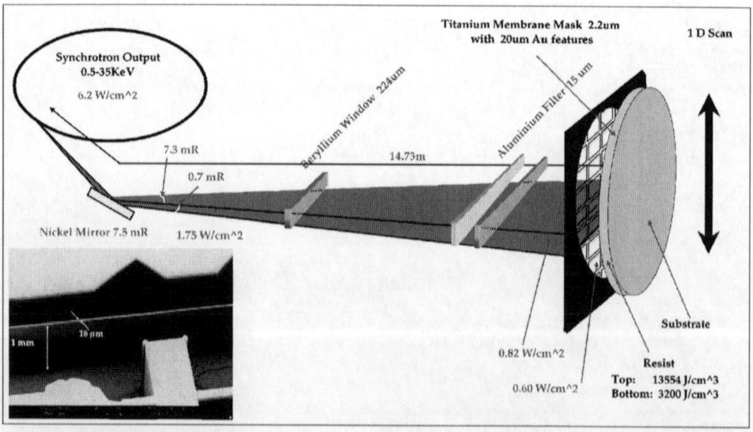

Figure 4.12 Schematic of an X-ray exposure station using a synchrotron source.

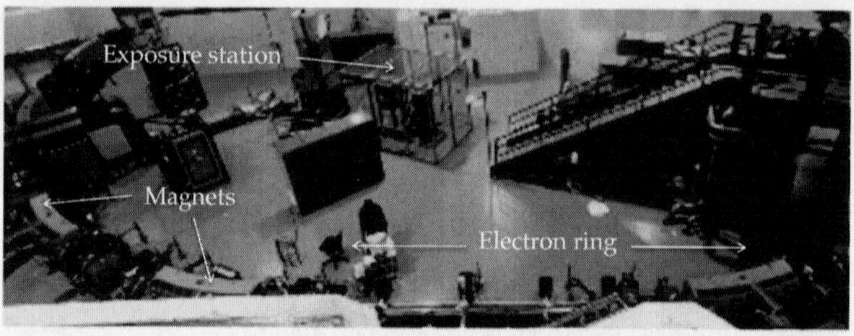

Figure 4.13 A synchrotron exposure station (courtesy Karlsruhe Institute of Technology). Note the circular ring of magnets forming the electron ring and the exposure station.

station as a rectangular beam whose length is at least the diameter of the wafer and width is approximately 1 cm. To cover the wafer area, the mask and the wafer must be scanned up and down until the resist is fully exposed. Figure 4.13 shows the basic features of an X-ray station in the experimental area of a synchrotron.

The exposure time can be determined for any synchrotron and beam line from a simplified relationship calculated from RALBEAM or similar programs. The estimated throughput of a LIGA installation for PMMA and SU-8 resists is shown in Fig. 4.14 as a function of

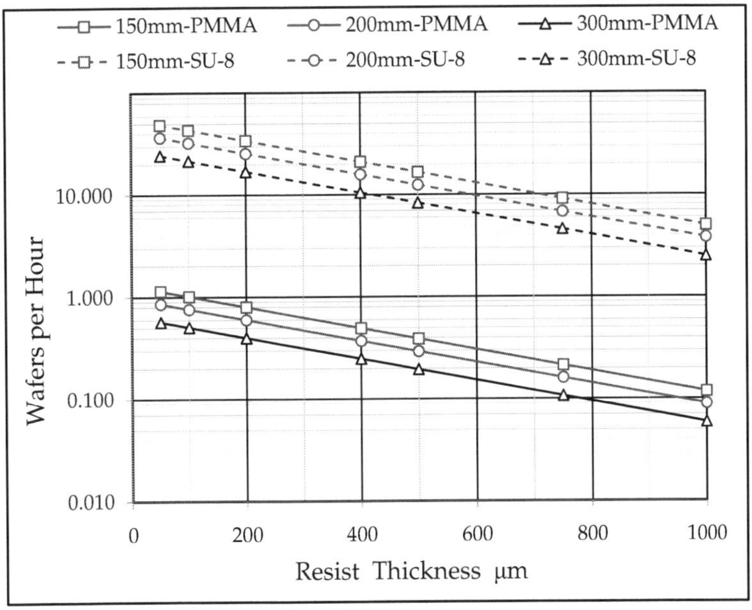

Figure 4.14 X-ray synchrotron throughput for PMMA and an SU-8 resist as a function of the resist thickness.

the resist depth. The following formula gives approximations for the exposure time and throughput (excluding load/unload overheads) for PMMA and SU-8 resists and forms the basis of the X-ray synchrotron simulator in Table 4.4.

For a PMMA resist the exposure time (T_{XPMMA}) is

$$T_{\mathrm{XPPM}} = 3727e^{0.0024H}\, W/I \tag{4.6}$$

where:

W = wafer diameter in mm
I = synchrotron current in mA
H = thickness of the resist in μm
T = exposure time (seconds)

and the throughput (R_{XPMMA}) in wafers per hour is

$$R_{\mathrm{XPMMA}} = 0.97e^{-0.0024H}\, I/W \tag{4.7}$$

For an SU-8 resist the exposure time (T_{XSU8}) is given by:

$$T_{\mathrm{XSU8}} = 88e^{0.0024H}\, W/I \tag{4.8}$$

Table 4.4 X-ray synchrotron simulator for PMMA and an SU-8 resist

Wafer diameter	200		mm
Resist thickness	100		μm
Synchrotron	ANKA (Karlsruhe)		
Beam current	120		mA
Resist type	PMMA	SU-8	
Exposure time T_{E*}	2,658	30	seconds
Load/unload	60	60	seconds
Total time	2,918	90	seconds
Batches/hour	1.3	40	
Batch size	1	1	wafer
Wafers/hour	1.3	40	

and the throughput (R_{XSU8}):

$$R_{XSU8} = 41e^{-0.0024}I/W \quad (4.9)$$

Studies have been made to determine the commercial viability of LIGA, particularly on cost (Lawes, 2009; Lawes, 2010) and convenience of use (Meyer et al., 2008). Despite the ability of X-ray LIGA to produce the highest resolution and quality of wall angle (Griffiths, 2004; Lawes, 2010) and its ability to pattern microstructures to a depth of several centimetres, the technique has not been widely exploited for commercial operation. This has been due partly to the perceived cost and the difficulty in accessing synchrotrons in research laboratories, particularly if attempting to integrate LIGA with a complementary-metal-oxide semiconductor (CMOS) process in a production foundry.

The company Microworks GmbH was spun out of the Karlsruhe Institute of Technology in 2007 to manufacture parts for watches and more recently specialist components for scientific instruments (Schultz, 2011). The main obstacle to the wider use of X-ray LIGA has been the development, for most MEMS applications, of more cost-effective and foundry-friendly techniques, such as DRIE (Section 4.2) and UV LIGA (Section 4.5).

NB: The load/unload time may be greater if the synchrotron station is difficult to access.

4.5 UV LIGA

It was soon realised that the SU-8 resist was sensitive to UV radiation at a wavelength of 365 nm with a relatively low absorption coefficient. This enabled HAR microstructure moulds to be exposed using commercially available chrome-on-quartz masks (Lee et al., 1995) and a conventional mask aligner. This has given rise to a cost-competitive technique, commonly known as UV LIGA, which is ideal to manufacture resist moulds for MEMS applications not only for research and development (R&D) applications but also for industrial production (Fu & Huang, 2006; Cui & Lawes, 2007).

The physical processes and manufacturing performance of UV LIGA (Chuang et al., 2002; Zhang et al., 2004; Liu et al., 2004) are now reasonably well-understood, although HAR structures cannot be fabricated with the same HAR that is possible with X-ray LIGA due to the effects of diffraction (Lawes, 2005).

A schematic of a wafer aligner is shown in Fig. 4.15. This is identical in most respects to the version shown in Chapter 3,

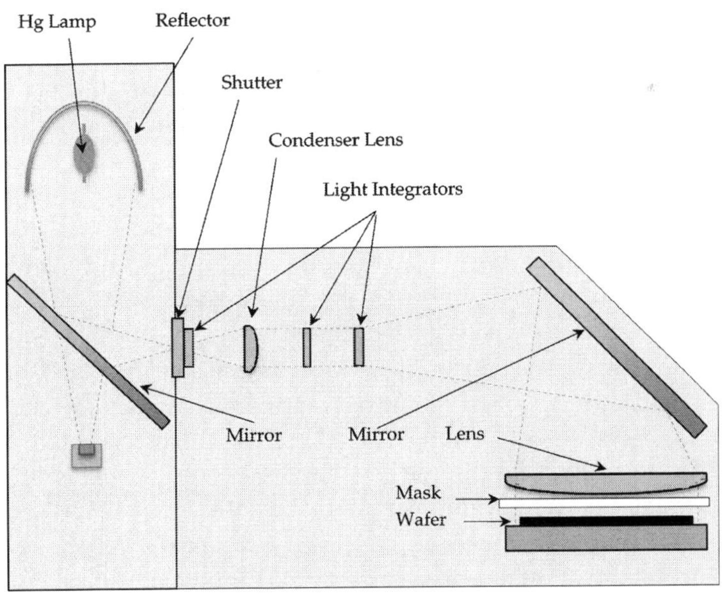

Figure 4.15 Schematic of a mask aligner used for UV LIGA.

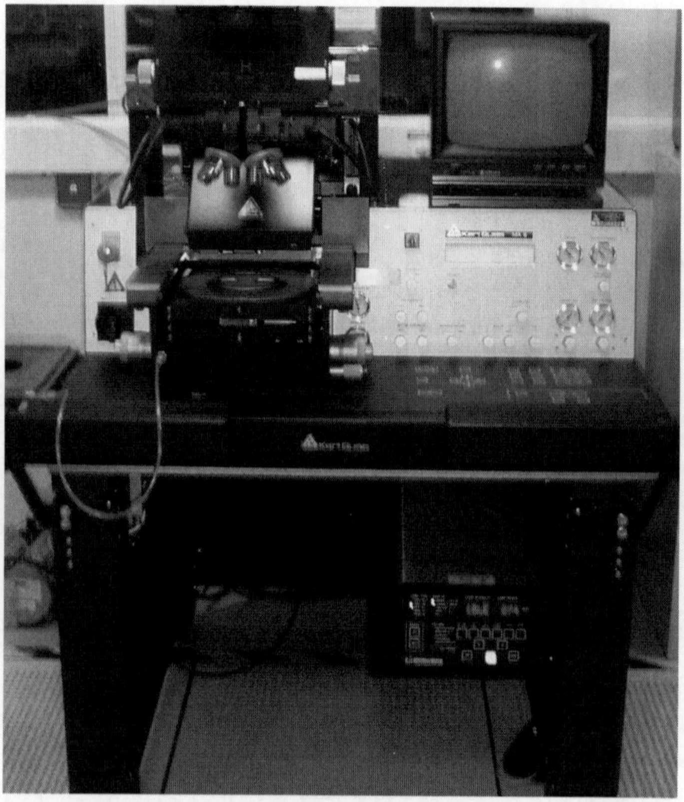

Figure 4.16 Suss MA6 UV aligner (author's archive).

Fig. 3.3, except that sequential exposure of both sides of the wafer is available. A typical commercial machine, suitable for UV LIGA is shown in Fig. 4.16.

The exact throughput with UV LIGA depends on the minimum radiation dose required at the bottom of the resist (1.5 Jcm^{-3} for SU-8) and the dose required for sufficient resist exposure to produce near-vertical microstructure walls (Reznikova et al., 2005). Assuming the dose at the bottom of the resist is required to be 5 Jcm^{-3}, then the exposure time (T_{365}) and the wafers per hour (R_{365}) are given by Eq. 4.10 and Eq. 4.11, respectively. The mask aligner is assumed to produce 6 $mWcm^{-2}$ of illumination, which, limited by an i-line filter to a 360–370 nm wavelength, produces a power density at the top of a wafer of 1.26 $mWcm^{-2}$.

$$T_{365} = 137e^{0.0029H} \tag{4.10}$$

$$R_{365} = 26e^{-0.00029H} \tag{4.11}$$

If the required resolution can be relaxed (say to 4–5 μm), then a wider spectrum of the mercury lamp may be used (e.g., 350–450 nm) so that the power density at the top of a wafer is increased to 5.1 $mWcm^{-2}$. Then the exposure time (T_{tot}) can be reduced (Eq. 4.12) and the throughput (R_{tot}) increased (Eq. 4.13).

$$T_{tot} = 87e^{0.0026H} \tag{4.12}$$

$$R_{tot} = 41e^{-0.0026H} \tag{4.13}$$

The exposure time and throughput for both spectra range is shown in Fig. 4.17 as a function of the resist thickness (Lawes, 2005), based on the above equations but including alignment and load/unload times. The UV LIGA simulator used in MEMSCOST is shown in Table 4.5.

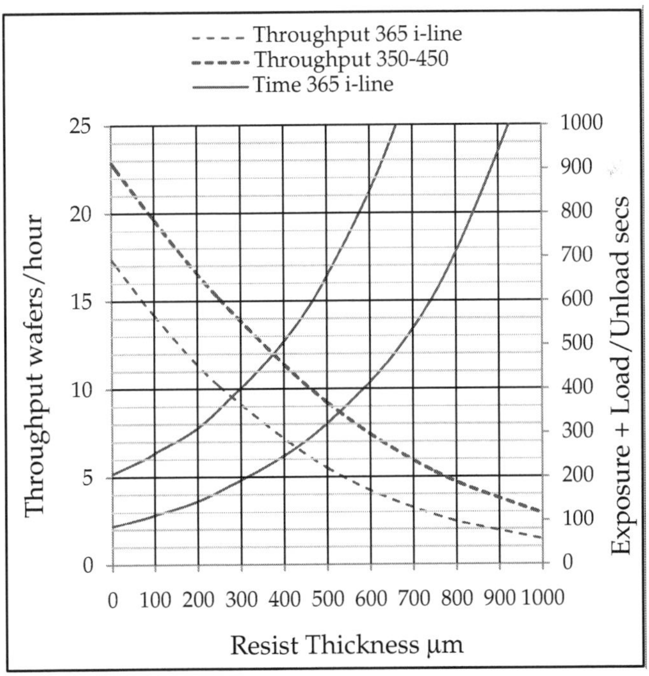

Figure 4.17 UV LIGA wafers per hour as a function of the resist thickness.

Table 4.5 UV LIGA simulator

Resist	SU-8	
Resist thickness	100	μm
i-line		
Energy density	6	mJ/cm^2
Dose at bottom of structure	5	J/cm^3
Registration	10	seconds
Exposure time	159	seconds
Alignment	10	seconds
Load/unload	60	seconds
Total time/batch	229	seconds
Batch size	1	
Batches/hour	15.9	

4.6 Bulk Micromachining

Bulk micromachining was one of the earliest methods used to manufacture relatively deep structures in silicon and is still used for simple applications (Kovacs et al., 1998). Bulk micromachining exploits the different etch rates for different etchants, where the etch direction depends on the crystallographic plane. The common etchants for silicon micromachining are solutions of potassium hydroxide (KOH) and tetramethyl ammonium hydroxide (TMAH). The etch rates for these etchants depend on the crystallographic planes and the temperature and are listed in Table 4.6 (the MEMSCOST bulk micromachining simulator). Ethylendiamine pyrocatechol (EDP) has been used in the past but may be carcinogenic and has been discontinued.

An example showing how bulk micromachining can be used to produce simple, accurate and deep microstructures is shown in Fig. 4.18. A silicon wafer orientated to the <100> plane is patterned with a masking material (e.g., a resist) in the form of a cantilever. Using an anisotropic etch, such as KOH, the etchant will continue in the <100> plane but slows down by two orders of magnitude at the <111> plane. The crystallographic angle between the <100> and <111> planes will define the sidewall angle at 54.74°.

Bulk micromachining relies upon the wet-chemical etching of materials, such as silicon, with preferential crystallographic planes

Table 4.6 Bulk micromachining simulator

Silicon			Etchant	Etch rate and plane orientation		
Depth	100	μm	Etch		μm/minute	
Plane	<100>			<100>	<110>	<111>
KOH			KOH @ 70°C	0.800	1.45	0.005
TMAH			TMAH @ 80°C	0.603	1.114	0.017
Rate μm/minute	0.8	μm/minute		0.8	1.45	0.005
Etch time	125	minutes				
Load/unload	1	minutes				
Total time	126	minutes				
Batches/hour	0.48					
Batch size	25	wafers				
Wafers/hour	12					

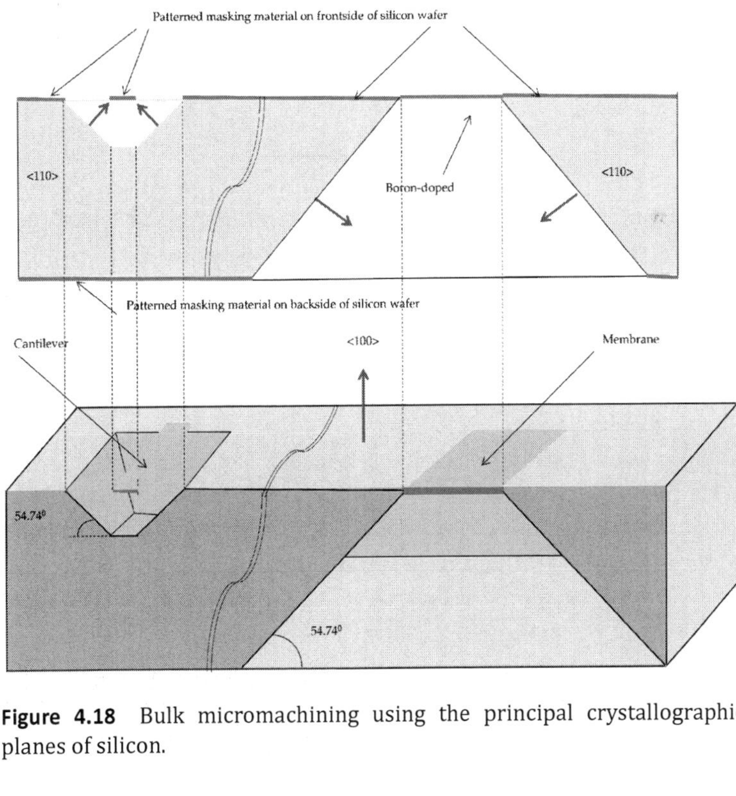

Figure 4.18 Bulk micromachining using the principal crystallographic planes of silicon.

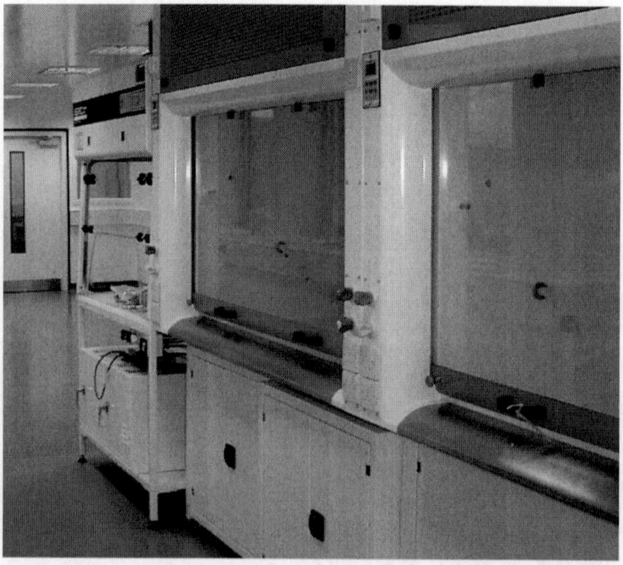

Figure 4.19 Wet bench suitable for bulk micromachining (courtesy Imperial College).

(Elwenspoek & Jansen 1998) and can produce relatively simple devices at very low cost. The main limitations are inflexibility, due to only a few crystallagraphic planes being available, and low resolution (50–100 μm). Cost of ownership is low as the equipment required is simple mask aligners, resist-handling tools and wet-etching baths with appropriate facilities for handling the toxic chemicals (e.g., see Fig. 4.19).

Bulk micromachining is still used in simple applications but has been mainly superseded by surface microengineering, which offers far more flexible design possibilities (see Chapter 3). A specialist application of bulk micromachining is the formation of suspended membranes. The required area of the membrane is patterned with a mask and boron-doped to form a thin layer of a few microns' thickness that is impervious to the etchant. The backside of the wafer is then patterned with a mask and the etchant introduced to the wafer. The silicon is etched away but stops when the boron-doped silicon layer is reached.

4.7 Wafer Bonding

Wafer bonding (Schmidt, 1998) is used to join two or more wafers together. It uses pressure and high temperature in various combinations and is probably the cheapest way in which two structured layers, typically silicon and/or glass, can be bonded together. The top and bottom wafers are typically aligned optically to an accuracy of 2–5 μm. Wafer bonding is particularly important for bonding a capping layer to a MEMS device (see Chapters 9 and 10) and to produce structures such as microfluidic channels, pressure sensors and micro-valves. Techniques are available to bond conducting and non-conducting substrates and are used not only to produce MEMS devices but also light-emitting diode (LED) and power devices. Schematics of the different bonding mechanisms are shown in Fig. 4.20.

Anodic bonding (Fig. 4.20a) is widely used for conducting substrates and relies on electron and ion conduction, for example, when bonding silicon and glass. No intermediate layers are required, but a voltage of 300–1,500 V is needed. Direct or fusion bonding (Fig. 4.20b) relies on van der Waals forces through intimate contact with ultra-flat surfaces (e.g., +/−0.5 nm). Room temperature annealing, followed by elevated temperature annealing, for example, >700–800°C, produces a strong bond without an intermediate layer.

Bonding with metal can be achieved by eutectic bonding (Fig. 4.20c) where the metal-silicon phase (363°C for Au-Si) is used. Slightly rougher surfaces can be bonded by a variety of solders, such as Pb-Sn at 183°C, Au-Sn at 282°C and Au-In at 156°C. Thermocompression (Fig. 4.20d) requires a force up to 100 kN and can bond rough surfaces using a relatively soft metal, for example, Au, as an intermediate layer. Eutectic bonding produces good hermetic seals at low temperatures and is being more widely used as device dimensions shrink.

Insulating substrates require an intermediate layer to bond successfully. The most widely used layer is glass frit where screen-printed glass frit can be deposited on one substrate; the second substrate is aligned, and then force (up to 3 kN) and temperature (350–450°C) are applied (Fig. 4.20e). Alternate intermediate layers

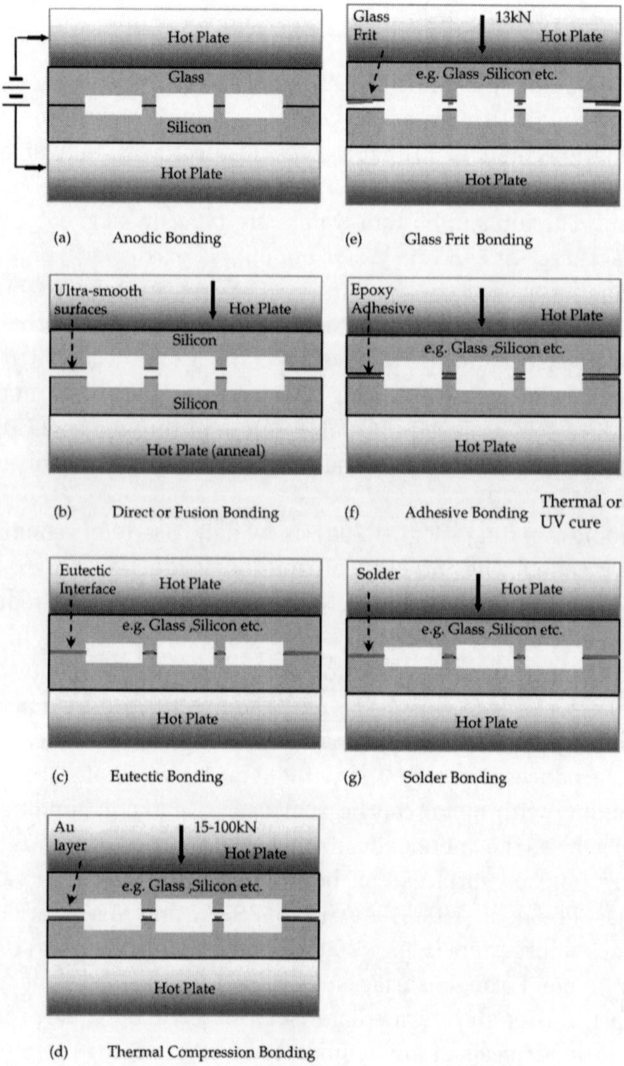

Figure 4.20 Wafer-bonding mechanisms.

are adhesives, such as epoxy glues and various polymers (Fig. 4.20f) and solder (Fig. 4.20g).

An interesting development with wafer-to-wafer (W2W) bonding is the pre-preparation of the wafer containing the microstruc-

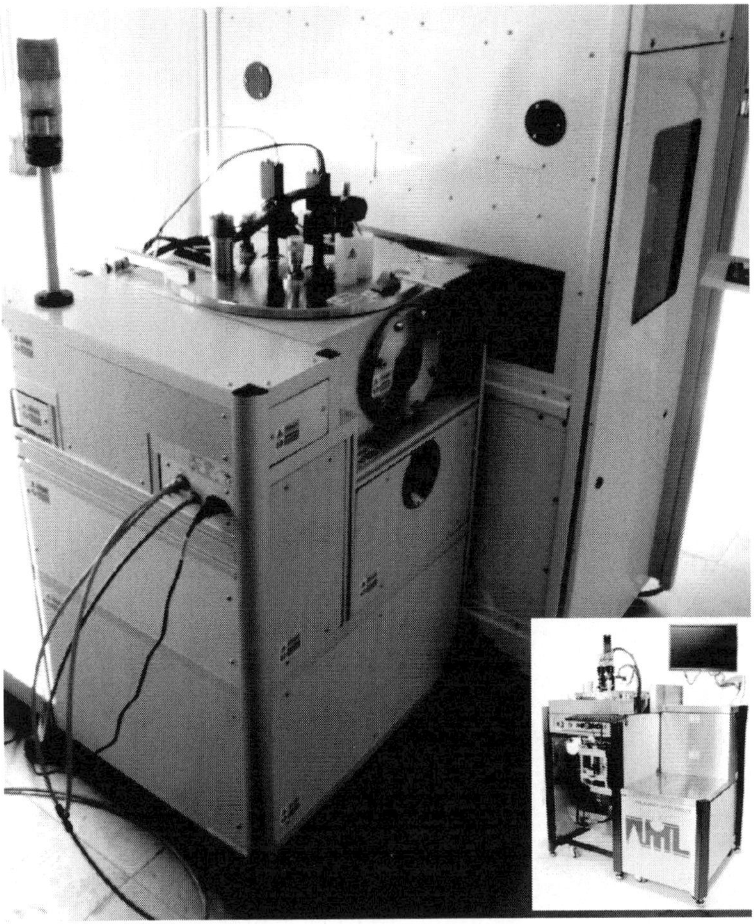

Figure 4.21 An auto-load/unload wafer-bonding machine with an R&D model shown as an inset (courtesy Applied Microengineering Ltd.).

tures (chips) to be bonded. Known-good-dies, (i.e., dies previously confirmed as working) from a variety of sources and sizes are pre-assembled onto temporary wafers and a chip-to-wafer (C2W) bond formed.

Commercial R&D and production machines are available (EVG, 2011), and some equipment manufacturers offer both machines and a specialist bonding service (AML, 2012). Figure 4.21 shows both a production and an R&D machine.

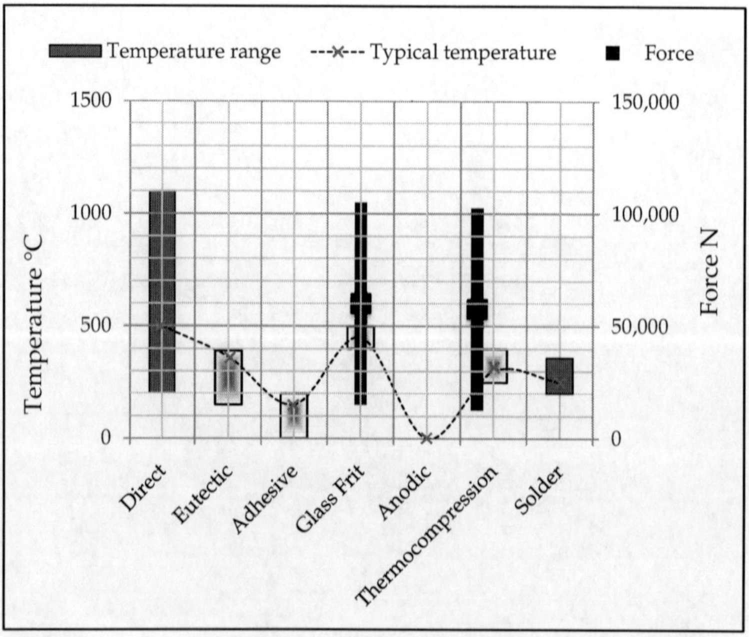

Figure 4.22 Temperature and pressure conditions for wafer bonding. The shaded columns indicate that an intermediate layer is used.

The wafer-bonding process used will depend on the materials to be bonded, the temperature required and, in the case of glass frit and a thermocompression technique, the force required to complete the bond. The different conditions are shown in Fig. 4.22. MEMSCOST has a wafer-bonding simulator that estimates the time taken and the throughput for each bonding method (Table 4.7).

4.8 Electrodeposition

Electrodeposition is capable of depositing stress-free metals at depths up to 500 μm. There are two forms of electrodeposition—electroplating, where the substrate on which material is to be deposited acts as an electrode, and electroless plating, in which deposition occurs without the need for an electric field (Schartzacher, 2006).

Table 4.7 Wafer-bonding simulator (includes load/unload time)

Wafer-bonding method	Hours/bond	Bonds/hour
Anodic	0.33	3.03
Anodic	0.33	3.03
Adhesive temperature	0.25	4.00
Adhesive UV-cured	0.15	6.67
Glass frit	2	0.50
Polymer	0.417	2.40
Fusion/direct	0.834	1.20
Eutectic	1.251	0.80
Library		

Electroplating requires the substrate to be immersed in an electrolyte and a voltage applied between the substrate and a second electrode. Electroplating requires electrically conducting materials. Electroless plating only requires a chemical solution from which metal ions are released spontaneously (Schacham-Diamond, 2003).

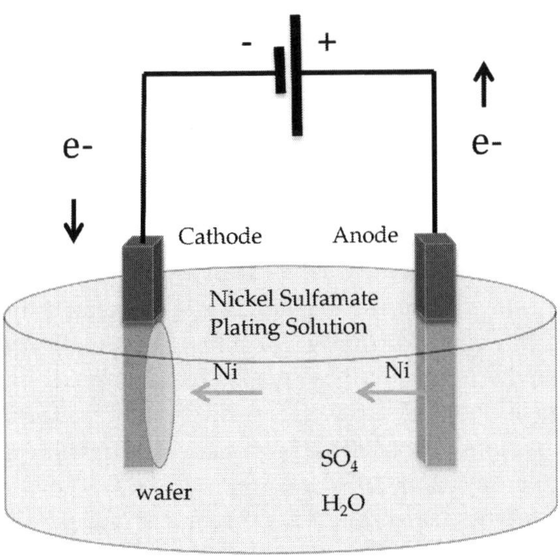

Figure 4.23 Schematic of an electrodeposition bath.

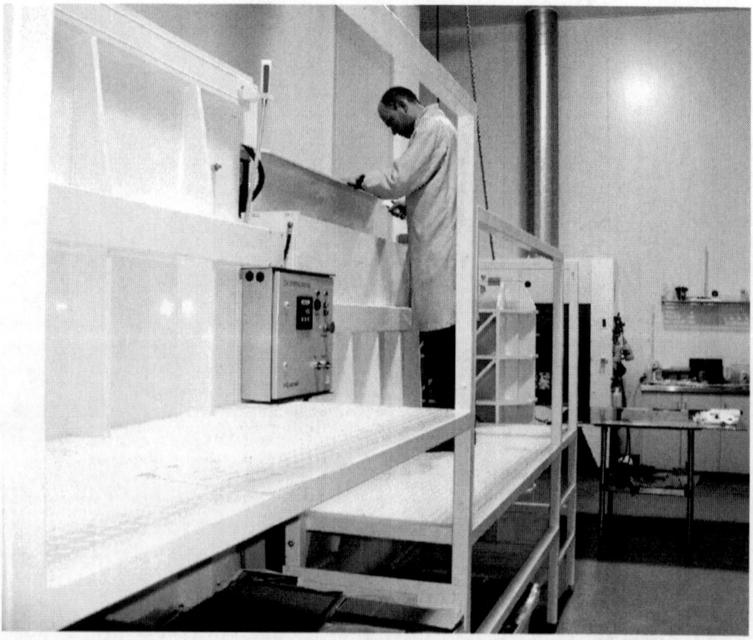

Figure 4.24 High-throughput electrodeposition equipment manufactured by Digital Matrix (courtesy MiniFAB, Australia).

An electrodeposition machine consists of an anode and cathode immersed in an electrolyte (a suitably conductive aqueous solution of metal salts) with a direct current flowing between the electrodes, as shown in Figs. 4.23 and 4.24. For example, electrodeposition of nickel uses a nickel anode immersed in a bath of nickel sulfamate ($NiSO_4{\cdot}6H_2O$) with a current density of 100–500 amps/m^2.

During the deposition process, positive nickel ions react with two electrons and are converted to metallic nickel at the wafer surface. At the anode, metallic nickel is dissolved into the electrolyte to replace those ions deposited at the cathode. The deposition rate can be calculated from a modified version of Faraday's law (amount of nickel deposited is proportional to current and time).

$$z/t = (I/A)(A_w m_p)/(e^- v)/\rho \tag{4.14}$$

where:

z = average nickel thickness (kg)
t = deposition time (h)
A_w = atomic weight of metal
v = valence of metal
ρ = density of metal (kg·m^{-3})
m_p = proton rest mass = 1.6725×10^{-27} (kg)
e^- = electronic charge = 1.6021×10^{-19} coulombs

From Eq. 4.14, substituting the fixed parameters

$$z/t = 37.4(I/A)(A_w/v)/\rho \quad (4.15)$$

where z/t = microns/hour.

The anode dissolution of metal ions balances those deposited on the cathode/wafer so that the anode efficiency (ion replaced/ion lost) is 100%. However, some small percentage of the current is used to dissociate the H_2O, thus reducing the cathode efficiency to 92–97%. Thus, for better accuracy, Eq. 4.15 should be multiplied by the cathode efficiency. The deposition rate for a variety of materials, calculated from Eq. 4.15, is summarised in Table 4.8.

To get high-quality deposits, with good adhesion to all parts of a microstructure, the current density is often ramped up from

Table 4.8 Deposition rates for some electrodeposited metals normalised to 100 A/m^2

Material	Atomic weight	Density kg/m^3	Valence	Rate 100 A/m^2 μm/hour/m^2	Electrolyte
Aluminium	26.9	2,700	3	12.6	
Chromium	51.96	7,100	2	13.8	
Copper	63.5	8,960	2	13.4	Copper sulphate
Gold	197	1,930	3	128.5	Gold chloride
Hafnium	178.5	1,310	4	128.6	
Nickel	**58.69**	**8,900**	**2**	**12.4**	Nickel sulphamate
Silver	107.9	1,050	2	194	
Tantalum	180.9	1,660	2	205.8	
Titanium	47.8	4,500	2	20.1	
Tungsten	183.85	19,300	2	18.0	
Zinc	65.38	7,140	2	17.3	

Table 4.9 Electrodeposition simulator for nickel

Parameter			
Nickel			
Wafer size	200		
Thickness	100	μm	
Deposition rate	12.4	μm/hour	Thickness
Ram amps	hours	μm/hour	μm
10	0.5	1.24	0.620
50	0.5	6.2	4.34
100	0.5	12.4	14.26
200	1.0	24.8	48.98
400	1.03	49.6	100.07
Sub-total	3.53	hours/batch	
Load/unload	1	minute	
Total time	3.55	hours/batch	
Batches/hour	0.28		
Deposition area	914	609	mm
Wafers/batch	4	3	12
Throughput	3.36	wafers/hour	

a low current density to the preferred high-throughput value. The MEMSCOST electrodeposition simulator for nickel is shown in Table 4.9. While the deposition rate is comparatively slow, electrodeposition has the advantage of being able to cover a relatively large area, thus allowing batch processing to depths of hundreds of microns.

4.9 Critical-Point-Drying

A major problem for many MEMS mechanical structures is the likelihood of stiction during the drying process. Stiction can cause damage to delicate components and lead to malfunctioning of the device. Damage is caused as the liquid crosses the liquid–gas boundary (Meinhart et al., 2006).

The solution, now widely adopted, is to dry the microstructures in carbon dioxide (CO_2) at its supercritical phase (Jaffri, 1999; Resnik & Clews, 2001). This transitional fluid directly releases

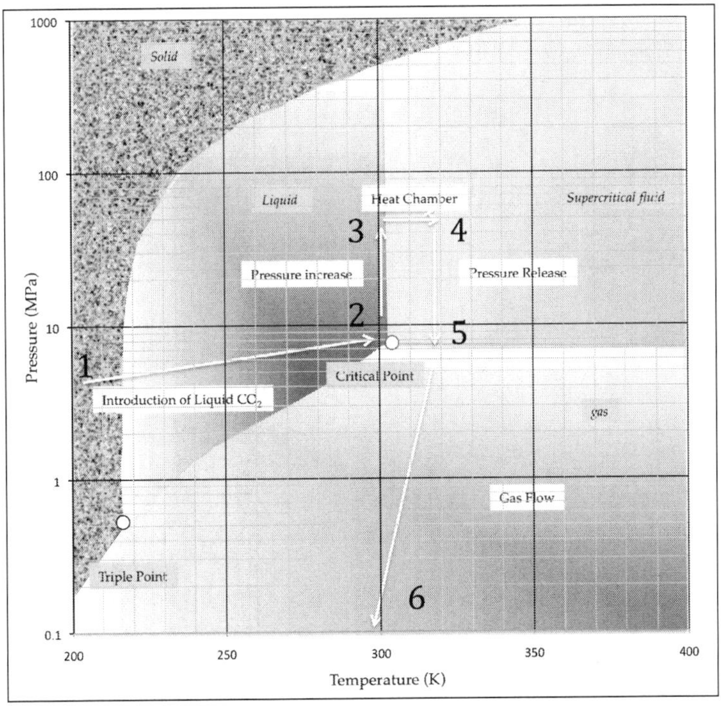

Figure 4.25 Phase diagram for carbon dioxide.

cleansing gases, without a liquid–gas interchange. In the supercritical phase liquid and gas have the same density and hence there are no surface tension effects. The sequence of process steps is shown in Fig. 4.25, the phase diagram for carbon dioxide.

Liquid carbon dioxide is introduced into the chamber of the critical-point dryer at ambient temperature and pressure (step 1) and the temperature and pressure increased until the critical point is reached (step 2). The pressure is raised (step 3) and the temperature increased beyond 32°C (step 4), the conditions in which the carbon dioxide is a super-critical fluid. Pressure is then reduced so that the carbon dioxide crosses the super-critical fluid–gas boundary (step 6) and begins the drying process. The whole process cycle takes approximately 60 minutes (see Table 4.10, the critical-point simulator). A commercially available machine to control the sequence of events is shown in Fig. 4.26.

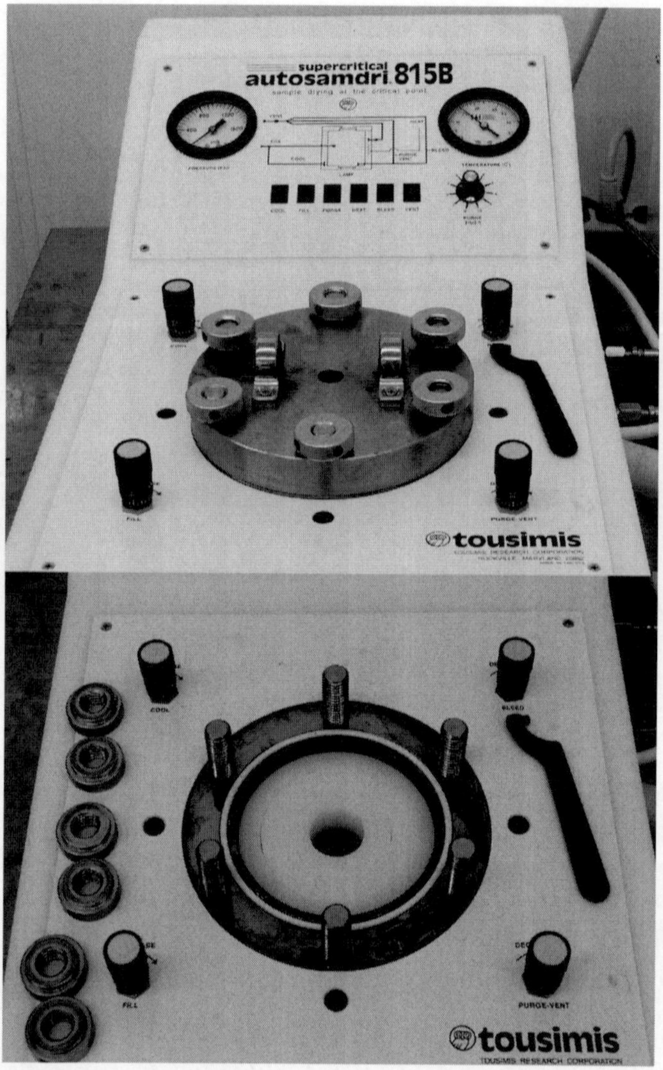

Figure 4.26 A single-wafer/multiple-die critical-point dryer (courtesy Scottish Microelectronics Centre).

Table 4.10 Critical-point dryer simulator

Critical-point dryer		Dimension
Wafer size	200	mm
Chamber diameter	216	mm
CP cycle time	59	minutes
Load/unload	1	minutes
Total time	60	minutes
Wafers/batch	1	
Batches/hour	1.0	
Wafers/per	1.0	

Abbreviation: CP, critical point.

4.10 Injection Moulding

Injection moulding is an inexpensive technique for large-scale replication of plastic parts. The equipment consists of three units: an injection unit, a moulding unit and a clamping unit. The key parts of an injection moulding machine are shown in Fig. 4.27 (also see Fig. 4.28).

The injection unit contains a hopper into which granules of the appropriate plastic (e.g., polycarbonate, polystyrene) are gravity-

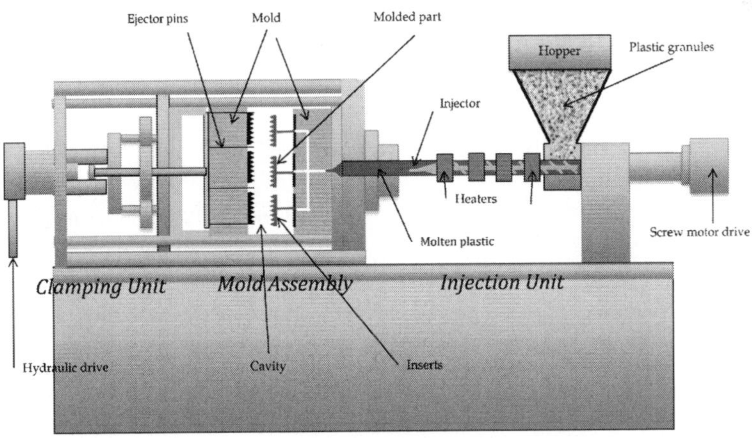

Figure 4.27 Schematic of an injection moulding machine.

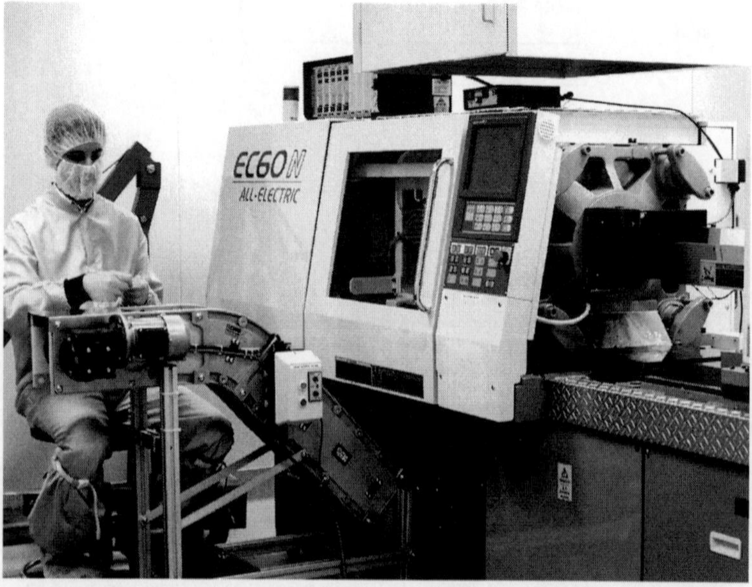

Figure 4.28 Toshiba EC60N injection moulding machine (courtesy mini-FAB, Australia).

fed into a heated barrel, which then liquefies the plastic granules. Formation of the plastic device starts when the clamping mechanism closes the two halves of the mould, leaving a cavity in the shape of the required device. The screw drive of the injection mechanism forces the liquefied plastic into this cavity under high pressure. After a short cooling period the clamping unit separates the mould and the device from one half of the mould. Ejector pins force the device from the other half of the mould. The equipment is then ready to repeat the cycle. The time to complete a cycle is short (10 seconds), leading to high throughput.

Injection moulding is particularly suitable for manufacturing plastic microfluidic parts where the device may be relatively large (e.g., several centimetres per side) and the channel widths approximately 50 μm. With a typical injection moulding machine the mould can hold a number of inserts that contain the design of the device. Therefore inserts for difference devices can be assembled into one mould, or the mould can contain a number of identical

Table 4.11 Injection moulding simulator

Injection moulder		
EC60N machine	100,000	$
Mould	22,000	$
Inserts/1 million injections	1,500	$
Inserts/mould	8	
Cost of basic machine	123,500	$
Cycle time inc. load/unload	10	seconds
Devices/hour	2,880	

inserts. In the latter case, the throughput for a given device can be further increased. The injection moulding simulator (Table 4.11) shows a machine with 8 identical inserts that can achieve a throughput of nearly 9,000 plastic devices per hour. MEMSCOST calculations show that devices in large volume, for example, over 5–10 million per year, can be manufactured for 1–2 cents each.

4.11 High-Aspect-Ratio Equipment Summary

HAR microfabrication technologies operate by different physical mechanisms, offer a different quality of micromachining (such as wall angle and roughness), expose resists at significantly different speeds and may machine material directly. Some techniques are more suitable for, or only applicable to, certain materials. A general schematic of the various HAR microengineering techniques is shown in Fig. 4.29. The major features affecting the cost of manufacture are discussed below.

HAR processes such as DRIE and bulk micromachining can produce structures in the wafer layer itself, whereas X-ray LIGA and UV LIGA require a resist intermediary. Excimer laser machining might do either, depending on the material. All these processes require some form of mask to define the device's design features. The time to process a wafer layer to the required depth depends on the machine technology and the degree of automation in substrate handling. Where high resolution and alignment accuracy are required, a step-and-repeat method must be used. If they are

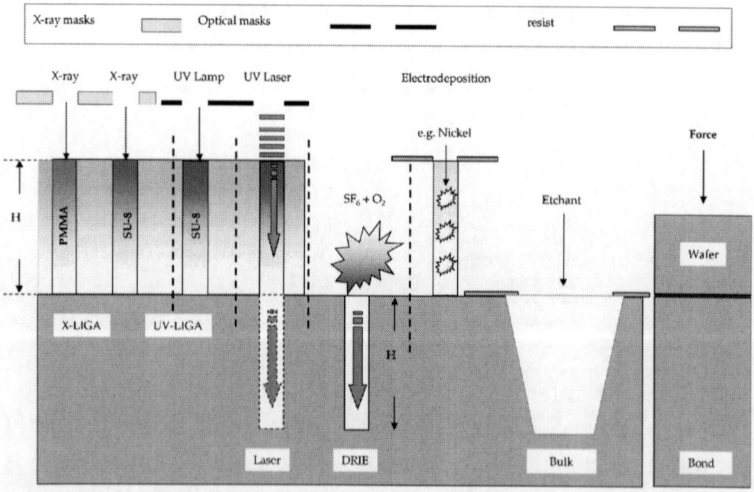

Figure 4.29 Summary of HAR techniques.

more relaxed, a whole wafer can be processed in one exposure. The exposure time algorithm depends on time taken for the exposing radiation to interact with either the resist or the substrate material.

The difference in resist thickness can be of the order of 500:1, and the resist-handling and development times will vary accordingly. The maximum layer thickness that each technique can machine is determined by the wall angle (which may be non-vertical by a few degrees) and the tolerances required for the top and bottom dimensions of the microstructure.

Table 4.12 shows the various aspects that make up the total time to process a wafer layer for the different technologies. A typical value of 100 μm film thickness has been chosen to demonstrate the throughput of the technology, taking into account the operation of the exposure machine and the associated resist technology. The maximum number of wafers per year is calculated, assuming that 7,560 hours per year are available after maintenance and unscheduled machine breakdown periods are deducted, which is effectively 90% of a foundry uptime (a total of 8,400 hours potential uptime is assumed).

Table 4.12 Typical data for HAR equipment and processes for 100 μm nominal film thickness

		Throughput	
HAR equipment and processes	$/unit	Wafers/hour	Maximum
DRIE	400,000	7.1	53,676
Laser	1,000,000	8.7	65,772
X-LIGA-P	1,200,000	1.3	9,828
X-LIGA-S	1,200,000	40	302,400
UV-LIGA-S	250,000	15.9	120,204
BULK	275,000	12	90,871
Wafer bond (anodic)	500,000	3	22,680
Electrodeposition (nickel)	350,000	3.4	25,704
Critical-point-drying	150,000	1	7,560

References

AML (2011). Bond centre wafer bonding machines and services. www.aml.co.uk.

Becker, E, Ehrfeld, W, et al. (1986). Fabrication of microstructures with high aspect ratio and great structural height, by synchrotron radiation, galvanforming and plastic molding (LIGA). *Microelectron. Eng.*, **4**(1), 35–56.

Chuang, Y-Y, et al. (2002). Reduction of diffraction effect of UV exposure on SU-8 negative thick photoresist by air gap. *Microsyst. Technol.*, **8**, 308–313.

Cui, Z, and Lawes, RA (1997). Low cost fabrication of micromechanical systems. *Microelectron. Eng.*, **35**, 389–392.

Davis, GM, Gibson, AF, Gower, MC, Lawes, RA, and Moody, RA (1983). Direct photoetching of resist materials using excimer lasers. *Microelectron. Eng. Cambridge*, 191–198.

Elwenspoek, M, and Jansen, H (1998). *Silicon Micromachining (Cambridge Studies in Semiconductor Physics and Microelectron. Eng.)*. ISBN: 0-521-59054-X(hb), 32–66.

EVG (2011). Wafer bonding systems. bonding.www.evg.com/en/products/bonding/wafer bonding.

Fu, C, and Huang, H (2006). Different methods for the fabrication of UV-LIGA molds using SU-8 with tapered de-molding angles. *Microsyst. Technol.*, **13**(3).

Goodall, FN, Lawes, RA, and Arthur, GG (1990). Deep UV optics for excimer laser systems. *Microelectron. Eng.*, **11**(1–4), 187–195.

Griffiths, SK (2004). Fundamental limitations of LIGAXray lithography: sidewall offset, slope and minimum feature size. *J. Micromech. Microeng.*, **14**(7), 999–1011.

Hafizovic, S, Hirai, Y, Tabata, O, and Korvink, JG (2003). X3-D: 3-D X-ray lithography and development simulation for MEMS. *Proc. 12th Int. Conf. Solid-State Sens. Actuators Microsyst., Boston, MA*, 1570–1573.

Harvey, E, Rumsby, PT, Gower, MC, and Remnant, J (1995). Microstructuring by excimer laser. *Proc. SPIE*, **2639**, 266.

Jaffri, IH, Busta, H, and Walsh, ST (1999). Critical point drying and cleaning for MEMS technology. *Proc. SPIE*, **3880**, 51.

Jain, K (1982). Ultrafast deep-UV lithography with excimer lasers. *IEE Electron. Device Lett.*, **EDL-3**, 53.

Jain, K (1990). *Excimer Laser Lithography*. SPIE Press, Bellingham, WA.

Kovacs, GT, Maluf, NI, and Peterson, KE (1998). Bulk micromachining of silicon. *Proc. IEEE*, **86**(8), 1516–1551.

Laermer, F, and Schilp, A (1994). Trench etch process for a single wafer RIE dry etch reactor. German Patent DE4241045C1 and US Patents 4855017 and 4784720.

Laermer, F, and Urban, A (2003). Challenges, developments and applications of silicon deep reactive ion etching. *Microelectron. Eng.*, **67–68**, 349–355.

Lawes, RA, Arthur, GG, Huq, E, and Schneider, A (2001). LIGA: a fabrication technology for industry? *Proc. SPIE: Design, Characterization and Packaging for MEMS and Microelectronics.*, **4593**, 145–155.

Lawes, RA (2005). Manufacturing tolerances for UV LIGA using SU-8 resist. *J. Micromech. Microeng.*, **15**, 2198–2203.

Lawes, RA (2007). Manufacturing costs for microstems/MEMS using high aspect ratio microfabrication techniques. *Microsyst. Technol.*, **13**(1), 85–95.

Lawes, RA (2009). Commercialization of LIGA. *Advanced Micro and Nano Systems*. Vol. 7, Chapter 7. Wiley-WCH.

Lawes, RA (2010). Linewidth design rules for microstructures, manufactured in PMMA resist using X-ray LIGA. *Microsyst. Technol.*, **16**, 411–421.

Lee, K, et al. (1995). Micromachining applications for a high resolution ultrathick photoresist. *J. Vac. Sci. Technol. B*, **13**, 3012–3016.

Lin, BJ (2009). *Optical Lithography.* SPIE Press, Bellingham, WA.

Liu, J, et al. (2004). Process research of high aspect ratio microstructure using SU-8 resist *Microsyst. Technol.*, **10**, 265–8.

Loechal, B, Goettert, J, and Desta, YM (2007). Direct LIGA service for prototyping: status report. *Microsyst. Technol.*, **13**, 327–334.

Meinhart, M, Miller, N, and Saif, MTA (2006). Dry stiction of microstructures—theory and experiment. *Proc. R. Soc. A*, **462**, 567–585.

Meyer, P, et al. (2002). A MS-Windows simulation tool for synchrotron X-ray exposure and subsequent development. *Microsyst. Technol.*, **9**, 104–108.

Meyer, P, Schulz, J, and Hahn, L (2003). DoseSim: MS-Windows graphical user interface for using synchrotron X-ray exposure and subsequent development in the LIGA process. *Rev. Sci. Instr.*, **74**, 1113–1119.

Meyer, P, Schulz, J, Hahn, L, and Saile, V (2008). Why will you use the deep X-ray LIGA technology to produce MEMS? *Microsyst. Technol.*, **14**, 1548–1557.

Resnick, PJ, and Clews, PJ (2001). Whole wafer critical point drying of MEMS devices. *Proc. SPIE*, **4558**, 189.

Reznikova, EF, Mohr, J, and Hein, H (2005). Deep photolithography characterization of SU-8 resist. *Microsyst. Technol.*, **11**, 282–291.

Rizvi, NH, Karnakis, D, and Gower, MC (2001). Micromachining of industrial materials with ultrafast lasers. *20th Int. Congress Appl. Lasers Electro-Opt.*, **92, 93**, *Congress Proc.*, 1511–1520.

Rumsby, PT, Gower, MC (1991). Excimer laser projector for microelectronics applications. *Proc. SPIE: Lasers Microelectron. Manuf.*, **1598**, 36–45.

Saile, V, et al. (2009). Commercialization of LIGA. *Advanced Micro and Nano Systems.* Vol. 7. Wiley-WCH.

Schmidt, MA (1998). Wafer-to-wafer bonding for microstructure formation. *Proc. IEEE*, **86**, 1575–1585.

Schultz, J (2011). Microworks—precision structures. www.micro-works.de

Schwarzacher, W (2006). Electrodeposition: a technology for the future. *Electrochem. Soc. Interface*, **15**(1), 32–35.

Schacham-Diamond, Y, et al. (2003). Electroless processes for micro- and nanoelectronics. *Electrochim. Acta*, **48**, 2987–2996.

Syms, R, Yeatman, E, Bright, V, and Whitesdes, G (2003). Surface tension-powered self-assembly of microstructures—the state-o-the-art. *J. Micromech. Syst.*, **12**(4), 387–417.

Tikare, V, et al. (2005). *The Role of Customized Computational Tools in Product Development.* Sandia National Laboratories Sand2005-2719.

Zhang, J, et al. (2004). Effect of exposure dose on the replication fidelity and profile of very high aspect ratio microchannels in SU-8. *Lab Chip,* **4**, 646–653.

Chapter 5

MEMS Testing

5.1 Introduction

A device can be tested at two levels during its manufacture. At the wafer level automatic test equipment (ATE) accesses and tests die while they are still an integral part of the wafer, that is, before packaging into individual devices. At the device level the ATE accesses the die after it has been packaged. While the test may be the same, the method of physically contacting the die is different depending on the test level. Wafers are accessed with precision probes from a 'probe card'. A robotic handler transfers packaged devices to an interface board, which has an assembly of sockets to take the packaged devices.

The demand for lower-cost and higher-performance ATE for integrated circuits (ICs) has produced advanced equipment that has been developed for all the main semiconductor products coming from the industry, including complementary-metal-oxide semiconductor (CMOS) memory, the central processing unit (CPU) and application-specific integrated circuits (ASICs). CMOS testing involves checking analogue parameters and logic signals for most products.

MEMS Cost Analysis: From Laboratory to Industry
Ron Lawes
Copyright © 2014 Pan Stanford Publishing Pte. Ltd.
ISBN 978-981-4411-06-6 (Hardcover), 978-981-4411-07-3 (eBook)
www.panstanford.com

This is not the case with microelectromechanical systems (MEMS) testing, where, depending on the device, electronic signals have to be measured in response to stimuli such as motion, pressure, temperature, fluids, sound and light. Testing of MEMS devices is still at a relatively embryonic stage, with much of the equipment and methodology confidential to companies (Beckford, 2009; Lemmers 2009).

Currently there is much more data on the design and performance of ATE for ICs than for MEMS. To estimate the cost of testing, the key components and performance of an IC ATE will be examined and then extended to a notional MEMS ATE system.

5.2 Testing of Semiconductors

The key components of ATE are:

- a base tester containing the components for signal detection and generation, for moving the wafer from die to die and for replacing the tested wafer by a new untested wafer at the end of the sequence.
- a probe card containing multiple tips designed to match the input/output (I/O) pad layout on a particular die or package. To increase throughput, high-performance systems have multiple heads, a process referred to as parallelism.

A schematic of such a system is shown in Fig. 5.1. During measurement each probe tip is briefly connected to an I/O pad on the die through a process known as 'touchdown'. The probe tips are in contact with the die pads for only a few seconds. The probe–die relationship is shown in Fig. 5.2. A probe and stage arrangement in a manual tester is shown in Fig. 5.3.

Equation 5.1 is a convenient algorithm to estimate the capital cost of an IC ATE system.

$$\mathrm{COST_{ATE}} = \mathrm{Cost_{BASE}} + \sum_{1}^{N} \mathrm{Cost_{pin}} \tag{5.1}$$

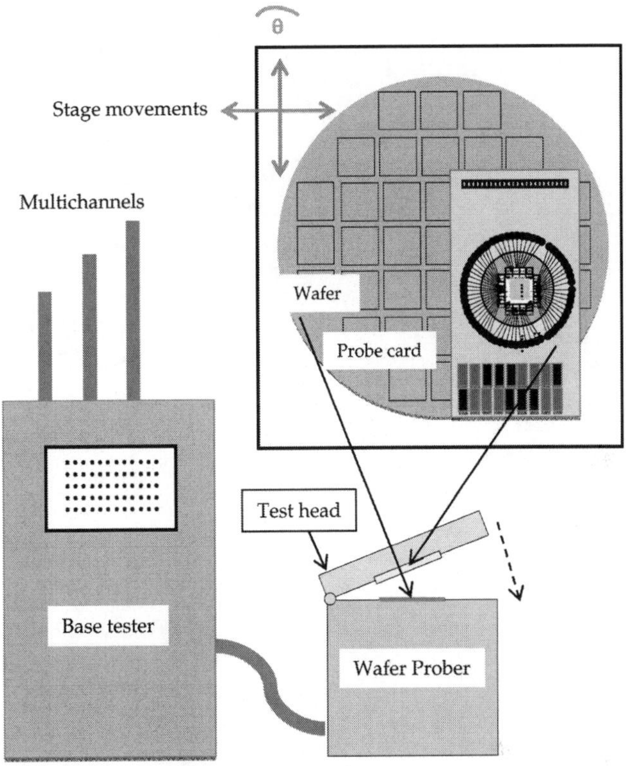

Figure 5.1 Schematic of ATE.

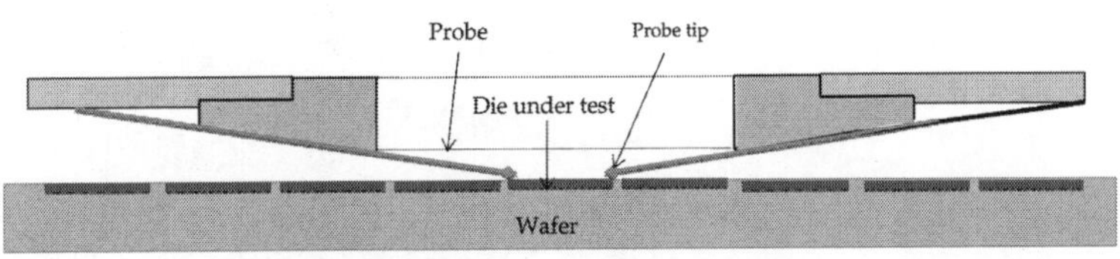

Figure 5.2 Schematic of a probe card showing contact with the die under test.

where:

$\text{Cost}_{\text{BASE}}$ = basic cost of a test system, excluding channels and probe cards

Cost_{Pin} = cost/pin

N = number of pins

Figure 5.3 Probe card (top) and a manual IC tester (bottom) (courtesy Scottish Microelectronics Centre).

Table 5.1 Estimated throughput per ASIC die (tested at the wafer level)

Wafer diameter	150	200	mm
Die size	1×1	1×1	
Untested die/wafer	13,576	24,584	
Test time per touchdown	4.5	4.5	seconds
Move-index to next die set	0.5	0.5	seconds
Time to test/die set	5.0	5.0	
Parallelism (die/touchdown)	8	8	
No. of touchdowns	1,697	3,073	
Time/wafer	8,485	15,365	seconds
ATE throughput	0.625	0.625	die/sec
	5,760	**5,760**	die/hour
Wafers/hour	0.42	0.23	

NB: The International Technology Roadmap for Semiconductors (ITRS, 2003) recommends costing by base tester + number of channels. For MEMS, as there are significant differences at the probe–die interface, MEMSCOST will use the earlier definition given in Eq. 5.1, that is, base tester + number of pins.

Table 5.1 gives the estimated throughput per ASIC die of a notional top-of-the-range tester. In this example the time to test at the wafer level is 8,485 and 15,365 seconds for 150 mm and 200 mm wafers respectively, taking into account a degree of parallelism, that is, multichannels. The overall throughput is 5,760 die per hour, that is, 0.42 (150 mm) wafers per hour and 0.23 (200 mm) wafers per hour.

It is now possible to estimate the cost of testing an ASIC, depending on assumptions about the time taken to test a die and the time taken to move to the next die(s). The load/unload time for a batch of wafers is assumed to be less than 0.5 seconds per wafer and will be ignored in the throughput calculation.

The data in Table 5.2 is derived from estimates given by the ITRS and updated by a modest increase in the overall ATE cost (estimated at 6% in total from 2003 to 2011). The data on the right of the table shows some typical component costs dependent on the type of IC being tested. It is assumed that the ATE is operated 24 hours per day, 360 days per year, at an overall uptime of 90%.

Table 5.2 Estimated cost per ASIC die (tested at the device level)

IC ATE				Base	Prober
				$	$/pin
ASIC			ASIC	583,879	7,431
C_{base}	583,879	583,879	Low-end ASIC	583,879	2,654
Cost/pin	7,431	7,431	Mixed signal	583,879	1,062
No. of pins	1,024	1,024	Commodity	583,879	31,848
Cost prober	7,609,344	7,609,344	Memory	583,879	1,062
Total ATE	8,193,223	8,193,223			
Depreciation (years)	5	5	2011	1.062 %	
Depreciation $	1,638,645	1,638,645			
Maintenance	1,638,645	1,638,645			
Operation	2,780,000	2,780,000			
	2,080,509	2,080,509			
Uptime %	90	90			
Total time/year	7,880	7,880	hours		
Test capacity	45,301,464	45,301,464	Die		
Wafer size	150	200	mm		
Cost/die	0.046	0.046	$/die		
Cost/wafer	625	1130	$		

The calculation estimates the annual cost of operating the ATE at $2,080,509, which would allow 45 million dies to be tested. MEMSCOST estimates a cost of 4.6 cents/die if the wafer is 100% tested and the equipment works at full capacity. The industry cost per die between 1984 and 2004 was approximately constant at 5 cents/die, rising to 9 cents/die from 1995 to 2001, before falling back to 4 cents/die by 2004. The cost per 150 mm wafer (13,576 dies) is estimated at $625, and the cost for a 200 mm wafer (24,584 dies) is estimated at $1,130.

5.3 Testing of MEMS

The technology of a MEMS tester will have some common features with those of an IC tester. MEMS ATE will have to measure the electrical signals resulting from a range of physical stimuli, depending on the device type (Maude et al., 2003; Tadigadama et al., 2003). Structurally, there are significantly fewer pins per MEMS die,

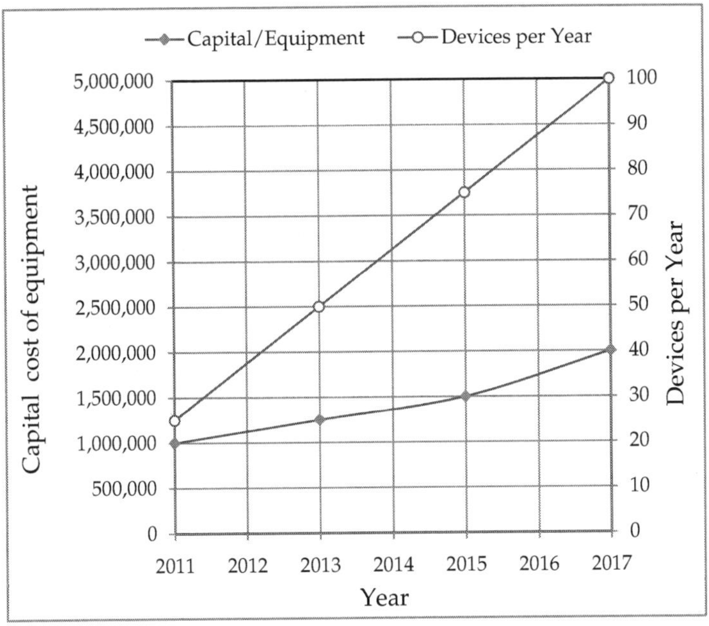

Figure 5.4 Estimated future MEMS test parameters (ITRS, 2011).

that is, 12–14, compared with 100–1,000 for an IC die (Williamson, 2011).

Guidelines for the cost and performance for future MEMS ATE systems, at wafer and device levels, have been set out by the ITRS for accelerometers (ITRS 2011 Table MEMS1), gyros (ITRS 2011 Table MEMS2) and microphones (ITRS 2011 Table MEMS3). Identical data for the capital cost of a MEMS tester has been forecast for the three products and is summarised in Fig. 5.4. The cost and performance of a wafer-level tester is predicted to be the same as a device-level tester, and this is the assumption by MEMSCOST when estimating the cost and throughput of MEMS testing.

From this data, the capital costs of a MEMS ATE tester can be calculated as a function of throughput in a given year. This is shown in Fig. 5.5 for device-level testing, along with the estimated parallelism to meet the performance requirement. For the MEMSCOST MEMS test model it is assumed that there is a fixed price for the basic MEMS tester ($791,000 in 2011) and a cost per device or

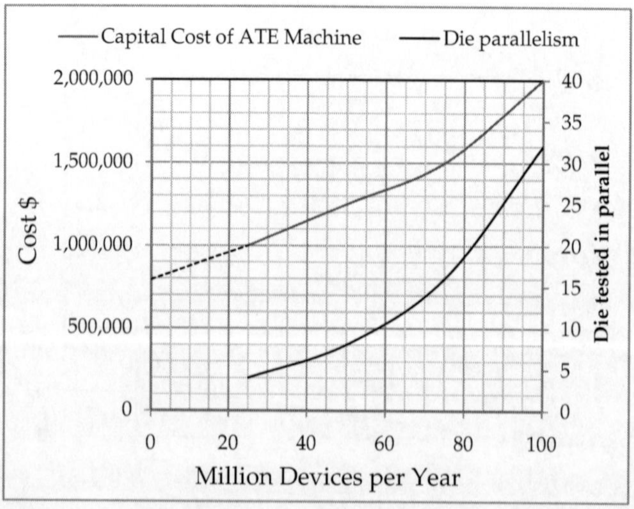

Figure 5.5 Estimated MEMS ATE capital cost as a function of installed test capacity.

die ($0.04 per device). In a year, such a tester would be capable of testing 25×106 devices or dies at a rate of 3,200 devices or dies per hour (assume 7,800 hours uptime). The MEMSCOST MEMS ATE Testing Simulator is shown in Table 5.3.

From the assumptions made, MEMSCOST estimates that the cost of testing a MEMS die is significantly lower than testing an IC die at $0.0195 per die compared with $0.046, respectively (see Tables 5.4 and 5.2).

Table 5.3 The MEMSCOST MEMS ATE Testing Simulator

Wafer diameter	150	200	mm
Die size	1 × 1	1 × 1	mm
Die per wafer	13,576	24,584	
ATE dies or devices per hour	3,200*	3,200*	
Time per die/device	1.1	1.1	seconds
Time per wafer	15,264	27,659	seconds
Throughput*	0.24	0.13	wafers/hour

*Throughput includes touchdowns, test time and load/unload.

Table 5.4 Estimated cost per MEMS die (tested at the device level)

Die size	1 × 1	mm
ATE base	791,000	$
ATE handler	209,000	$
Capital cost	1,000,000	$
Depreciation	200,000	$
Cost of operation	287,000	$
Total annual cost	487,000	$/year/hour
Capacity	25,000,000	Devices or dies per year
Cost/device or die	0.0195	$ per device or die

MEMS testing requires significant development if it is to keep up with the future reduction in fabrication costs. Test equipment can be reduced in cost with a move to standardisation, but this is limited in scope by the different stimuli and measurements needed for different products. The predicted increase in throughput, which is probably necessary to reduce costs, and the consequent reduction

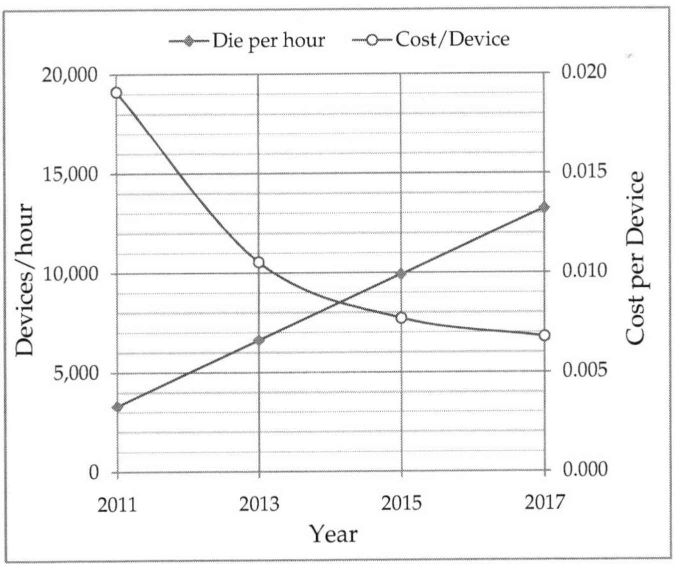

Figure 5.6 Future performance estimates for MEMS device and die testing.

in time per test are shown in Fig. 5.6 and is derived from the ITRS predictions.

References

Beckford, W, et al. (2009). Analog devices reduces MEMS test costs with PXI and LabView.sine.ni.com/cs/app/doc/p/id/cs-13005.

Electroiq (2011). MEMS testing units bolster Rasco product line with high parallelism.www.electroiq.com

ITRS (2003). Test and test equipment. *The International Technology Roadmap for Semiconductors*, Table 21.

ITRS (2011). Table MEMS 1-micro-electro-mechanical systems MEMS1 accelerometer requirements-testing. *The International Roadmap for Semiconductors.*

ITRS (2011). Table MEMS 2-micro-electro-mechanical systems MEMS1 accelerometer requirements-testing. *The International Roadmap for Semiconductors.*

ITRS (2011). Table MEMS 3-micro-electro-mechanical systems MEMS1 accelerometer requirements-testing. *The International Roadmap for Semiconductors.*

Lemmers, D (2009). ADI and NI develop low-cost MEMS test. *Semiconductor International*, 5 August 2009.

Maudie, T, et al. (2003). MEMS manufacturing testing: an accelerometer case study.*ICT Int. Test Conf. IEE*, 843–849.

Tadigadapa, S, and Najafi, N (2003). *Transactions of ASME*. Vol. 125, 816—823.

Williamson, J (2011). Automated test equipment: an overview of MEMS and non-MEMS switching options. *MEMS Investor J. April.*

Chapter 6

MEMS Packaging

6.1 Introduction

During the early development of commercial microelectromechanical systems (MEMS) products, the reduction of costs was inhibited by the wide variety of non-standard manufacturing processes, materials and packaging. In many cases, packaging was the dominant cost. In recent years, much of the MEMS industry has moved to using silicon as a material, with silicon processes and standard integrated circuit (IC)-like plastic packages (Gilleo, 2005). For example, MEMS accelerometers are silicon-based and typically use quad flat no. (QFN) (see Fig. 6.3) plastic packages. The QFN form is known as an open-cavity package and is completed by the bonding of a plastic lid to form a hermetic seal.

However, some major MEMS products, such as microfluidics, still require non-standard packaging (Evans, 2004). The packaging techniques adapted for MEMS from the IC industry will be used to develop a cost simulator for packaging and for MEMSCOST cost calculations.

It is helpful to review the sequence of steps used by the semiconductor industry. Individual dies must be separated from the wafer by a process known as *dicing* or *singulation* and eventually

MEMS Cost Analysis: From Laboratory to Industry
Ron Lawes
Copyright © 2014 Pan Stanford Publishing Pte. Ltd.
ISBN 978-981-4411-06-6 (Hardcover), 978-981-4411-07-3 (eBook)
www.panstanford.com

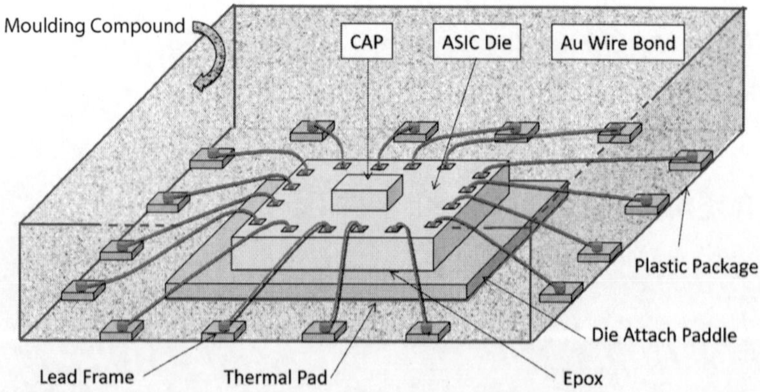

Figure 6.1 Schematic of a plastic package showing the key components.

assembled into the chosen *package*. A preparatory step requires the wafer to be mounted on a frame, and a tape is applied to the wafer to ensure the separated die is easy to handle. Dicing is done by sawing down the scribe lanes with a resin-bonded diamond wheel and separating the dies. The diced wafer is then cleaned by de-ionised water to prepare the individual dies for insertion into the selected package. The individual dies are then attached to the *Die Attach Pad*, which is part of the package, by an attach tool capable of pick-and-place to an accuracy of $+/-3$ μm. The attached die is wire-bonded from pads on the die to pads on the package to enable connections to systems external to the package. The packaging process is completed by filling the cavity with a moulding compound (mainly for ICs) and epoxy-glueing a lid to hermetically seal the packaged device.

A schematic of a packaged MEMS device is shown in Fig. 6.1. Note the presence of a cap, which is designed to protect moveable and vulnerable MEMS parts (e.g., accelerometers), and in many cases the cap also provides a controlled atmosphere. Wafer-level packaging is a method of capping a die (Esashi, 2008) and is achieved by bonding together two wafers, one containing the MEMS device and the other containing the cap (see Fig. 6.16). The combined wafers are subsequently diced (see Section 6.7).

Figure 6.2 A selection of packages developed by the semiconductor industry (courtesy Scottish Microelectronics Centre, Edinburgh).

6.2 Packages

Over the last few decades the number of packages developed for the semiconductor industry has run into the hundreds (the author has identified 288), with different package styles and with pin counts ranging from 8 to 256. For a sample of packages developed for the semiconductor industry see Fig. 6.2.

These packages are made from either ceramic or plastic. Ceramic is now mainly reserved for demanding environments as it is estimated to cost 10 times the cost of a suitable plastic material (Ashby, 2010).

Many of the available packages are unsuitable for MEMS, and the higher-pin-count packages are not currently required for MEMS devices, as 14–16 pins per package are adequate to conduct signals to the outside environment. MEMS uses plastic packages to keep overall costs to a minimum.

Prominent packages for accelerometers now tend to be either the 'QFN leads' series (e.g., QFN-14, QFN-16), where the pads are designed to enable direct surface-mounting to a printed circuit board, or the 'land grid array' (LGA) (e.g., LGA-10, LGA-14), which is essentially a 'ball grid array' package without the spherical balls

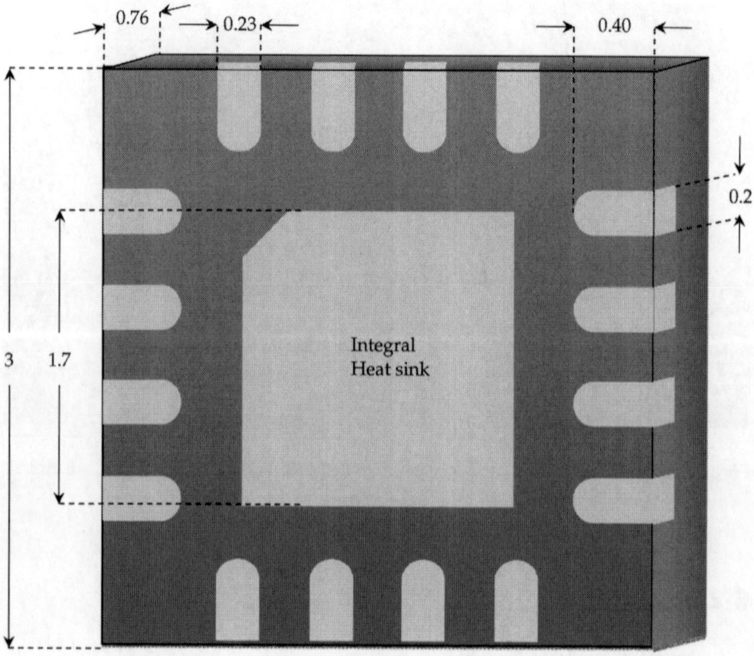

Figure 6.3 The QFN-16 lead plastic package (average dimensions in mm).

(Robinson, 2012). Both packages make provision for heat transfer to the printed circuit. The QFN package has an exposed thermal copper pad, and the LGA package relies on good thermal bonding to an external heat sink. A dimensioned schematic for the QFN-16 package is shown in Fig. 6.3.

The cost of a given package in MEMSCOST is notional, as the cost depends on the volume ordered, the supplier, etc., and is not generally available in the literature. As an alternative, a MEMSCOST-type calculation is available through a *Package Cost Simulator*. These calculations are based upon the cost of materials, such as plastic and various metals, the capital cost of equipment, estimated running costs and manpower.

The MEMSCOST *Package Cost Simulator* (Table 6.1) details the calculation for a QFN-12 plastic package, separately costing the plastic body, the individual copper/nickel gold-plated pads and the integral heat sink. This is a good example showing how cost

Table 6.1 The MEMSCOST *Package Cost Simulator* (shown for plastic QFN-12)

Package	QFN-12	mm	Volume		
L	4	0.25	External	14.4	mm^3
B	4	0.25	Internal	9.14	mm^3
T	0.9	0.25	Lid	4	mm^3
Density kg/m^3	1,330		Net volume	9.26	
Cost/m^3	4000	Cost $/kg	3.0	Cost ($)	3.7040E-05
Pads			Pins	12	
L	0.4		Total volume	0.0184	mm^3
B	0.23	0.001	Au volume	0.000436	mm^3
T	0.2		Cu volume	0.017964	mm^3
Metals	Gold	Copper	Al	Nickel	
Density kg/m^3	19,300	8,930	2700	8,900	
Cost $/kg	45,900	9.35	2.6	18	
Per pad	3.862E-04			2.878E-06	3.891E-04
Total pads	0.0046344			3.454E-05	4.669E-03
Heat sink					
L	2.15	mm^3	Volume	2.311	mm^3
B	2.15	mm^3			
T	0.5	mm^3		Cost	1.930E-04
Manufacture	Assembly	Pads	Heat sink	Total	
Capital	$	500,000	500,000	500,000	1,500,000
Depreciation	$	100,000	100,000	100,000	300,000
Maintenance	$	10,000	10,000	10,000	30,000
Manpower (maint.)	$	6,000	6,000	6,000	18,000
Fixed cost (m/c)	$	116,000	116,000	116,000	348,000
Packages/year		1.0E+08	1.0E+08	1.0E+08	
Packages per hour		7,500	10,000	10,000	
Machine-hours/year		13,333	10,000	10,000	33,333
No. of machines		2	2	2	6
Fixed cost	$	232,000	232,000	232,000	696,000
Manpower cost/hour	$	36.78	36.78	36.78	
Variable	$	490,388	367,800	367,800	1,225,988
Sub-total cost	$	722,388	599,800	599,800	1,921,988
Manufacture/package	$	0.00722	0.00600	0.00600	0.01922
Total cost	$	0.00726	0.01067	0.00619	0.02412
Total price (at 50% profit)				0.048	

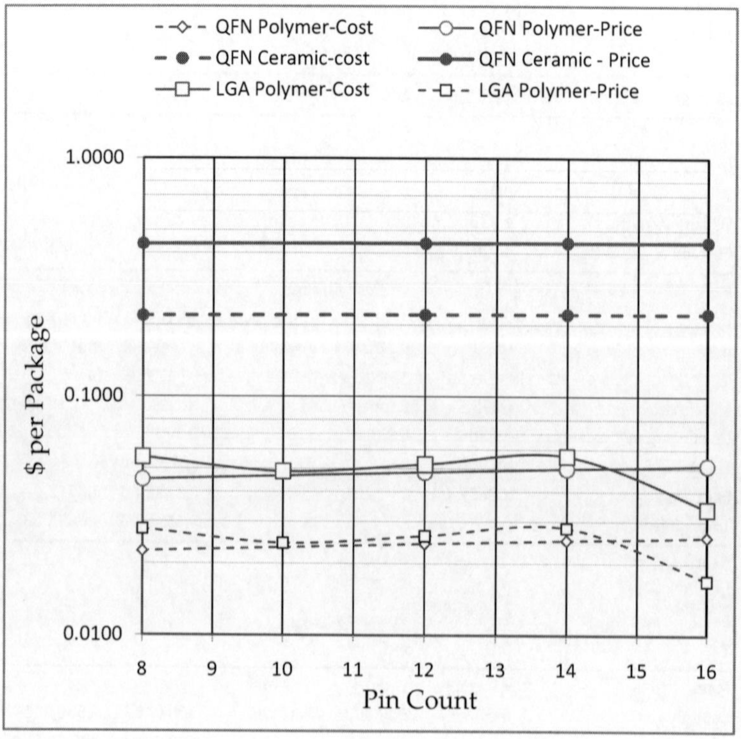

Figure 6.4 Estimated cost of manufacture and selling price for QFN and LGA plastic packages.

analysis can be used to give at least an order of magnitude (or better) estimate of the cost to manufacture any item.

Figure 6.4 estimates the cost to manufacture QFN and LGA packages in large quantities and the likely purchase price to a MEMS foundry (the profit margin of the package manufacturer is assumed to be 50%). Note that the variations in cost are reasonably proportional to pin count, except for some variation with LGA packages. This is partly due to the overall package dimensions not being monotonic with pin count.

An estimate is made for the same packages but replacing the polymer with ceramics. It is assumed that the processing of ceramics involves higher-capital-cost equipment (approximately four times) than with plastics and that the process cycle is longer (300 minutes compared with 20 minutes).

While the estimates are sensitive to pad and manufacturing throughput costs, the assumptions made confirm the estimates that a plastic package costs less than \$0.05 and a ceramic package is an order of magnitude more expensive at \$0.5. However, this data is only estimates, and for specific product/foundry cost analysis, actual data values should be used, where available.

6.3 Dicing

As noted in Chapter 2, Fig. 2.8, the dies are surrounded by scribe lanes of sufficient width (typically 50–100 μm) to enable the wafers to be converted into separate dies. This process is referred to as *singulation* or *wafer dicing* and, in the IC industry, is achieved by mechanical sawing. Figure 6.5 shows the saw-positioning encoder, the diamond- encrusted grinding wheel and the slurry dispenser.

Figure 6.5 The wafer-mounting stage, the saw head and the slurry-dispensing tube of a dicing machine (courtesy Loadpoint, UK).

An encoder can place the saw blade, within the scribe lane, to an accuracy of +/ − 4 μm.

Sawing MEMS wafers with a mechanical saw is potentially damaging to the delicate MEMS structures either from the production of debris from the diced silicon wafer or from the water used to cool the saw. Capping the die before dicing, that is, wafer-level packaging, alleviates the problem. An alternative method of dicing uses a laser as the cutting tool, and a new technology known as 'stealth laser cutting' is emerging (Lares, 2008). The stealth laser method of dicing offers faster cutting rates (300 mm/second), can produce less debris and can cut narrower scribe lanes (20 μm) due to the reduced size of the laser beam. However, the mechanical method is widely used in industry and will be used in MEMSCOST's machine simulation. A description of the method follows.

The wafers are mounted onto a dicing tape, which has a sticky backing that keeps the dies in their relative positions after the scribe lanes have been cut.

The total length of the scribe lanes that must be cut is shown diagrammatically in Fig. 6.6. After cutting all the scribe lanes in one direction, the wafer is rotated through 90° to cut the remaining

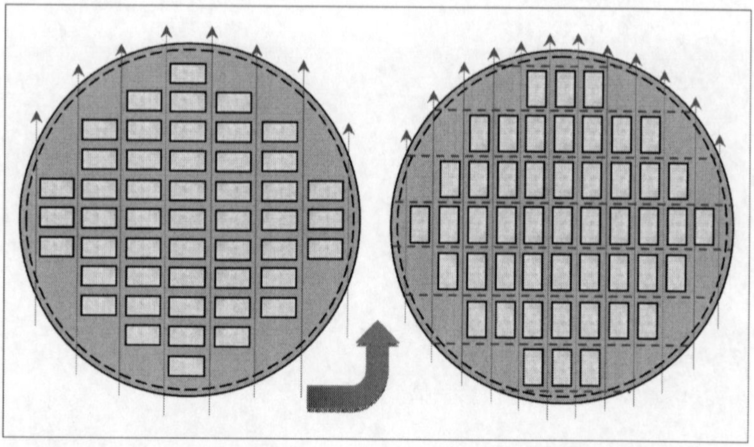

Figure 6.6 Dicing operation first cut (left) followed by rotation of die through 90° and a second cut to complete dicing.

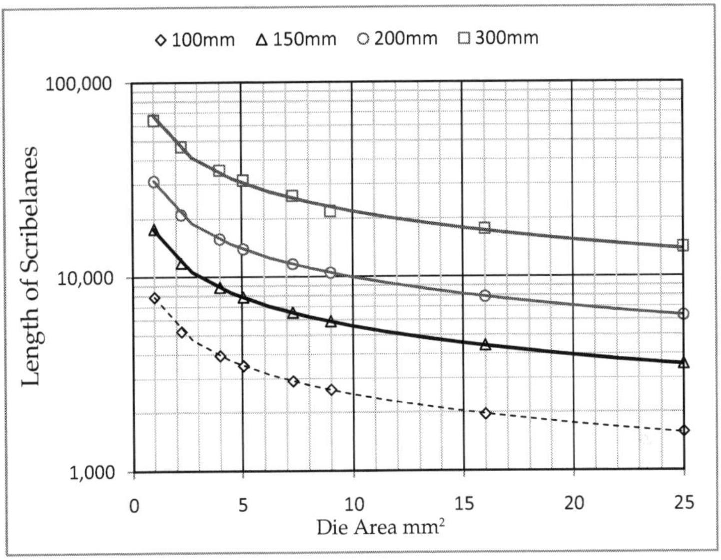

Figure 6.7 Total length of the substrate to be cut as a function of the wafer diameter and die area.

scribe lanes. An approximate formula may be derived for the total length of cut for given wafer diameter, die length and die width.

$$C_L = 0.8W^2/\sqrt{LB} \tag{6.1}$$

where:

C_L = total length of vertical and horizontal cuts (mm)
L = die length + scribe lane width (mm)
B = die width + scribe lane width (mm)
W = wafer diameter (mm)

Equation 6.1 can be used to calculate the total length of the scribe lanes that must be cut for a given die size on 100 mm, 150 mm, 200 mm and 300 mm wafers (Fig. 6.7). A *Dicing Simulator* can now be derived, where throughput is a function of wafer and die size and the cutting rate of the dicing saw (Table 6.2).

Table 6.2 Dicing Simulator

Wafer diameter		200	mm
Die length		10.0	mm
Die width (mm)		10.0	mm
X lanes (18)	*Y* lanes (18)	36	
Total length to cut (mm)-from Equation 6.3.1		3,148	mm
Cutting rate		50	mm/second
Cutting time		63.0	seconds
Turnaround time/cut		0.5	seconds
Total turnaround time		18	seconds
Load/unload time		60	seconds
Total time		141	seconds
Wafers/hour		25.5	

6.4 Die Attach

Die attach involves selecting a die from the diced wafer and placing it onto a package substrate or a metal lead frame (Maluf & Williams, 2004; Hsu, 2004; Lindros, 2010).

A good, separated die on the diced wafer is recognised by a pattern recognition system, picked up, aligned to the substrate and bonded to the package substrate using several different bonding techniques, for example,

- eutectic bonding, which requires a silicon–gold combination at relatively high temperatures (e.g., 450°C) in a nitrogen/hydrogen atmosphere to prevent oxidation of the lead frame
- soft solder bonding, where a heated solder blob (from a solder wire) is placed on the lead frame and the die placed on the hot solder
- flip chip bonding, which is essentially the same as soft solder bonding, except the die is flipped over so that solder bumps bond the die to the substrate (important for very high input/output [I/O])
- epoxy bonding, where epoxy, or other adhesives (such as a polymer), is dispensed on to the bond pads or substrate and the die placed on the substrate

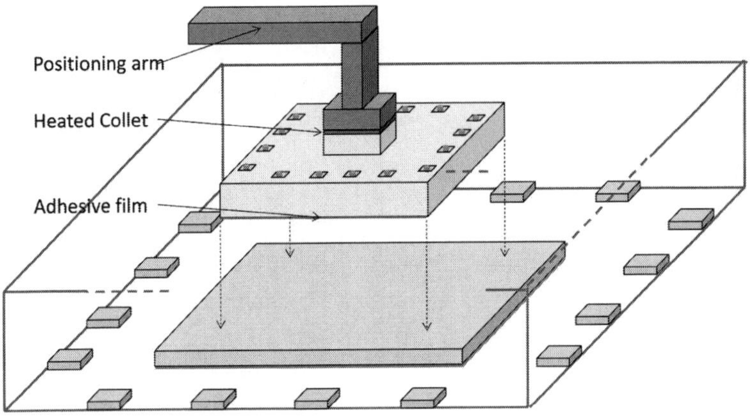

Figure 6.8 Die attach machine showing the positioning arm prior to attaching the die to the paddle.

Figure 6.8 shows a schematic showing the die attach arm positioning the die prior to final attachment to the package. The *Die Attach Simulator*, using an epoxy bond, is shown in Table 6.3.

Table 6.3 Die Attach Simulator

Die attach	MEMS	ASIC
Pick/place	10,000	10,000
Time/package	0.36	0.36
Dispense epoxy	0.04	0.04
Total time	0.4	0.4
Packages/hour	9,000	9,000
Materials		
Area	9	0
Thickness	0.1	0.1
Volume of epoxy	0.9	0
	9E-10	0
Cost $/litre	10,000	10,000
Cost $/die	0.000000009	0

Abbreviation: ASIC, application-specific integrated circuit.

6.5 Wire Bonding

Wire bonding is used to connect a die to the outside world. It connects pads on the die to pads on the package, via thin gold, aluminium or copper wires. Gold wire diameters can be as thin as 12.5 μm (typically 25 μm), copper from 75 μm to 250 μm and aluminium typically 100 μm.

The wires are bonded onto the pads by ultrasonic energy, with thermosonic or thermocompression mechanisms, forming either a ball-type or a wedge-type connection (Dixit, 2006). With ball bonding, the tip of the wire is melted (by capacitive discharge) and the ball is formed by surface tension and then fixed to the selected pad by ultrasonic energy. With wedge bonding, the wire is placed against the pad and ultrasonic energy applied to form the bond between the pad and the wire. With thermocompression bonding, pressure and vibration are used to form a solid-phase bond. With the addition of heating, the process is known as thermosonic.

The general layout of a single die with the bond pads and wire bonds is shown in Fig. 6.9. Figure 6.10 shows the commonly used mechanisms and bonding types for the three common wire materials. In the case of a multi-die assembly wire bonding also takes place from the MEMS die to the ASIC die.

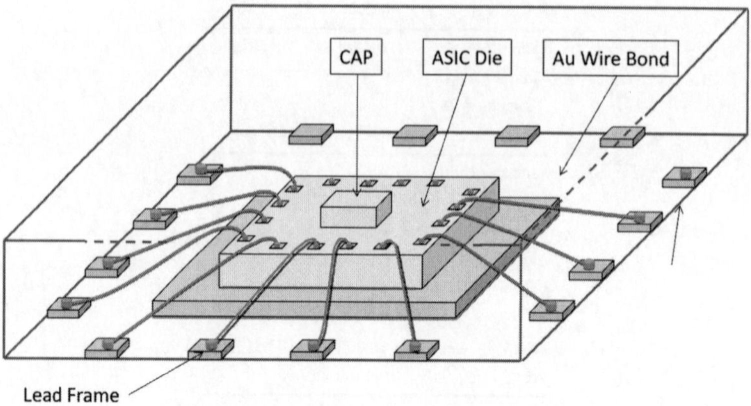

Figure 6.9 Wire bonds from die to lead frame.

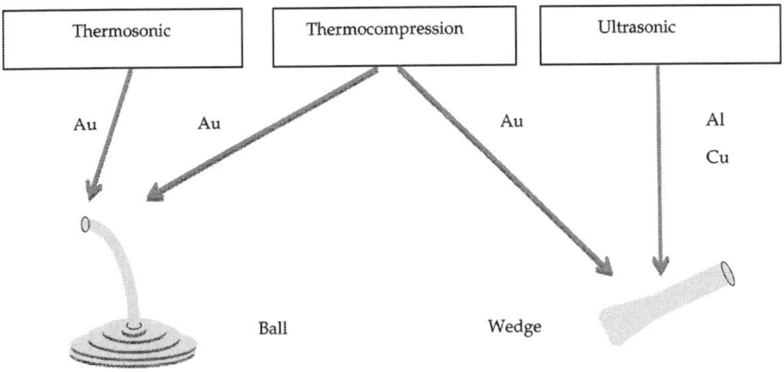

Figure 6.10 Mechanisms and bonding types for wire bonding.

Gold has been the standard material for wire bonding as copper had problems with oxidation and needed higher energies to create the bond. However, by 2011, the cost of gold had risen astronomically from previous levels and had reached \$45,900/kg compared with copper at \$9.35, a price difference of 5,000 times! There is now a move to replace gold wire with copper, wherever possible. The use of copper instead of gold can reduce packaging costs by 5–20%, depending on the diameter of the wire used and the pin count of the package.

The costs of wire bonding will depend on the type of wire used, the diameter of the wire and the length and shape of the wire loop between the ASIC/MEMS die and the lead frame pad. Figure 6.11 plots the cost of gold and copper wires as a function of the wire diameter for a 12-pin QFN package. The cost reduction achieved by replacing gold with copper is also shown (10% for 25.4 μm wires). The cost reductions with larger wire diameters would be even more significant. Figure 6.12 plots the cost of gold and copper wires as a function of the pin count, assuming 25.4 μm wires. The cost reduction achieved by replacing gold with copper is also shown (10% for 12 pins). The cost reductions with larger pin counts would be even more significant.

The packaging costs calculated in later chapters are based on the use of gold wires during the wire-bonding stage. A laboratory wafer-dicing and bonding machine is shown in Fig. 6.13. The MEMSCOST *Wire-Bonding Simulator* is shown in Table 6.4.

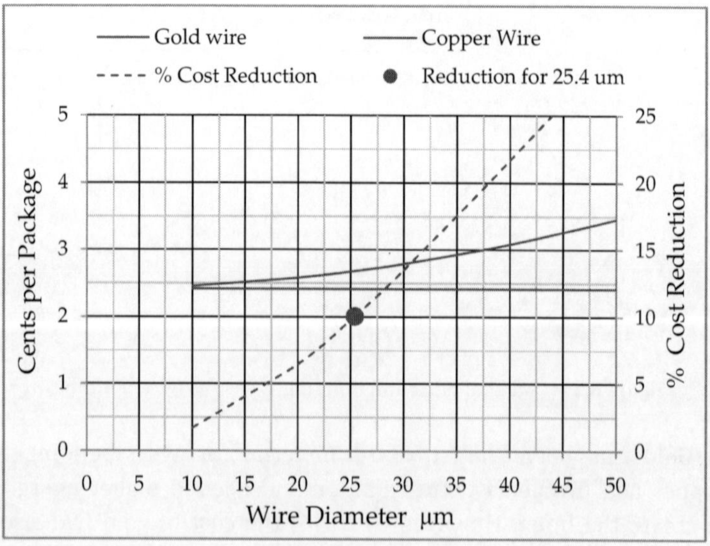

Figure 6.11 Packaging cost of gold and copper wire and % cost reduction as a function of the wire diameter for a QFN-12 package.

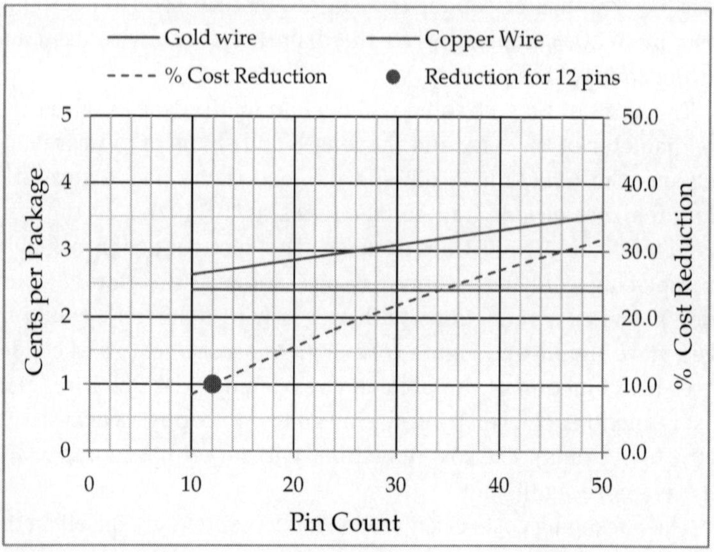

Figure 6.12 Packaging cost of gold and copper wire and % cost reduction as a function of the pin count for a QFN package with 25.4 μm wires.

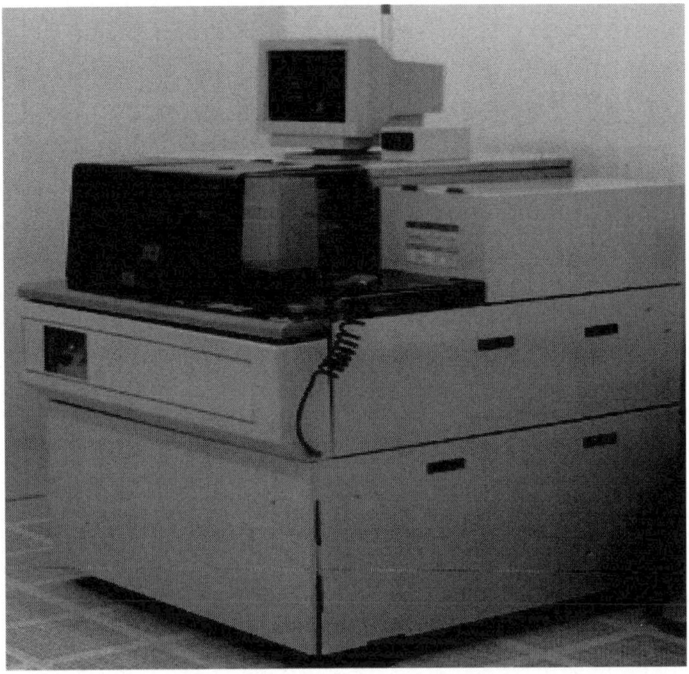

Figure 6.13 Wafer-dicing and bonding machine (courtesy Scottish Microelectronics Centre, Edinburgh).

Table 6.4 Wire-Bonding Simulator

Package	QFN-12		
Gold			
Wires/second	20		
Time/package	0.625	seconds	
Packages/second	1.67		
Packages/hour	**2,156**		
Materials	Wire	Density kg/m^3	19,300
Diameter (mm)	0.0254	Mass kg	4.73E-09
Length (mm)	0.5	Cost $/kg	45,900
Volume/mm^3	2.45E-04	Cost/wire $	2.17E-04
Volume m^3	2.45E-13	No. of wires	12
$/package			0.0026
Metals	Gold	Copper	Aluminium
Density kg/m^3	19,300	8,930	2,700
Cost $/kg	45,900	9.35	2.6

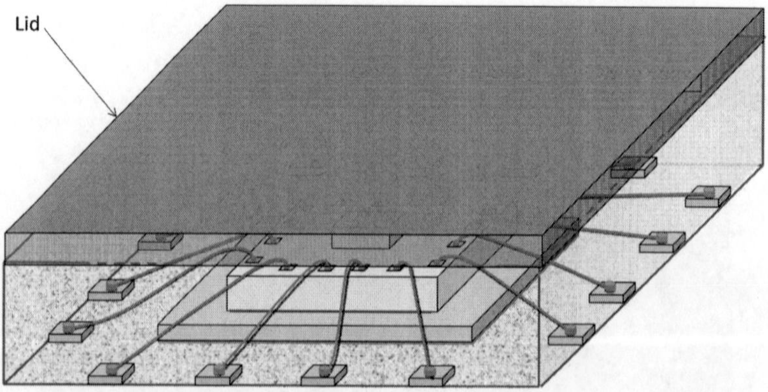

Figure 6.14 Package sealed with a plastic lid.

6.6 Lid Seal

The final packaging step is a sealing process, where a flat plastic lid is bonded to the package body once the die attachment and wire-bonding steps have been completed. This is achieved by glueing along the edge of the package using an automatic or semi-automatic machine. A schematic of the completed package is shown in Fig. 6.14. Throughput for a notional lid attach machine is shown in the MEMSCOST *Lid Attach Simulator* (Table 6.5).

With ASIC and MEMS chips the lid is a simple plastic, metal or ceramic fixture, but some MEMS devices need access to the outside environment. For example, MEMS devices such as pressure sensors and microphones would require an aperture through which pressure and acoustic stimuli, respectively, could pass. In the case of

Table 6.5 Lid Attach Simulator

Deposit glue	10	0.1	seconds
Place lid per hour	10,000	0.36	seconds
Load/unload per wafer	60	–	seconds
Total time		0.46	
Packages/second		2.2	
Packages/hour		7,920	

an optically sensitive device, the lid can be hermetically sealed but with a transparent window for line-of-sight to the sensitive area.

6.7 Single- and Multi-Die Packaging

Several architectures are used to package the MEMS and ASIC dies (Allan, 2004).

Figure 6.15a shows a single-die version of a MEMS device, where the MEMS component is an integral part of the die and is manufactured with the ASIC signal processor as part of complementary-metal-oxide semiconductor (CMOS) fabrication. Where a multi-chip approach has been adopted, only the MEMS die is capped and the ASIC and capped MEMS dies wire-bonded into a common package.

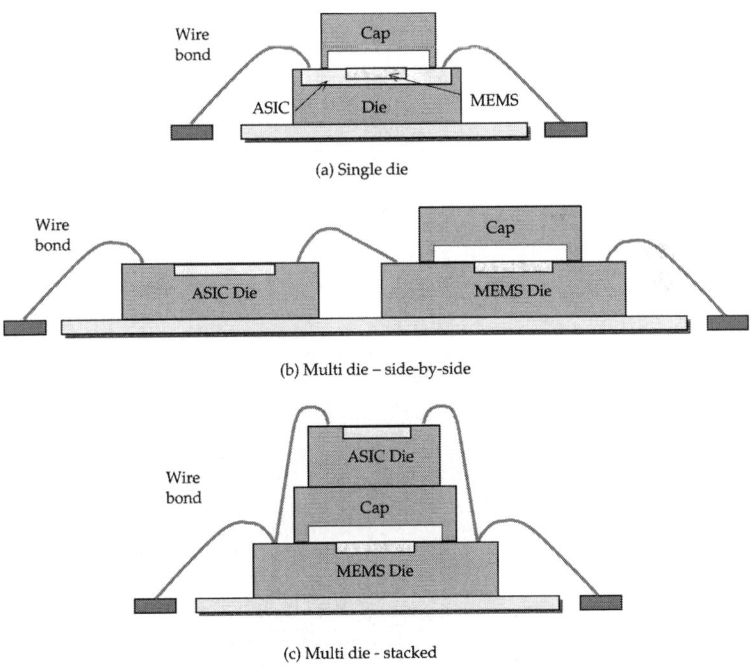

Figure 6.15 Packaging architectures for a MEMS microstructure and its ASIC signal-processing die. Compare (a) with Fig. 9.3, (b) with Fig. 9.4 and (c) with Fig. 9.6 in Chapter 9.

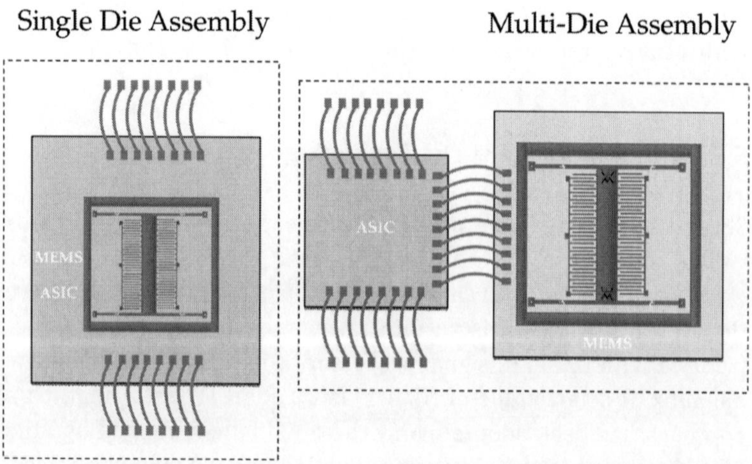

Figure 6.16 Schematic of single-die and multi-die assembly (not to scale).

This method of packaging is referred to as 'side by side' packaging (Fig. 6.15b). An alternative form of multi-die packaging is shown in Fig. 6.15c, where the ASIC die is mounted on the top of the capped MEMS die or vice versa. This method of packaging is referred to as 'stacked'.

The single-die assembly is the most efficient in terms of package space, wire bonding and a single die attach step. The multi-die approach requires more wire bonding and, in the case of side-by-side assembly, more package space (see the schematic in Fig. 6.16). The stacked version requires more vertical height in the package and/or wafer thinning.

From the available design data for accelerometer dimensions, a multi-die assembly ASIC die is 26% larger than the MEMS sensor on average (see Chapter 9, Table 9.2). In contrast, for a microphone a single-die assembly ASIC is 14% smaller than the MEMS die (see Chapter 10, Table 10.1). MEMSCOST takes these dimensional variations into account when calculating costs.

As previously noted, capping is an important step in protecting MEMS components. There are several methods of manufacturing the cap and attaching it to the die. An elegant solution is shown in Fig. 6.17. Steps 1 and 2 form a cavity in a silicon wafer using lithography and deep reactive-ion etching (DRIE). Glass frit is deposited around

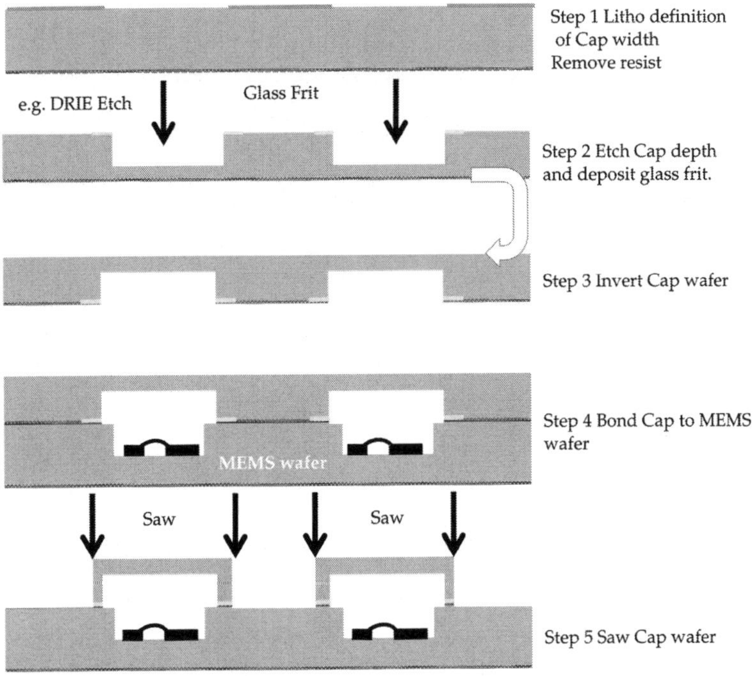

Figure 6.17 Formation of a protective cap for a MEMS die.

the edge of the cavity. The fabricated wafer is then inverted (step 3) and brought into contact with the wafer containing the MEMS die. In step 4 the two wafers are then bonded together eutectically (see Chapter 4, Section 4.7). The final step (step 5) is to saw the wafer to form the sides of the cap. The surplus silicon is removed from the bonded wafers.

6.8 MEMS Packaging Summary

MEMSCOST can be used to estimate the total cost of packaging a MEMS device for a given set of assumptions. In the example below the device is a MEMS accelerometer and assumes a 1 mm × 1 mm single die, a 200 mm wafer with an 85% yield, with further assumptions for the capital cost of equipment and throughput. A plastic QFN-12 package is assumed. The percentage of the

Table 6.6 Percentage packaging costs for a 1 mm × 1 mm die on a 200 mm wafer

Process	$	$/package	%	%
Add cap	267,219	0.0027	0.5	3.2
Dicing	135,035	0.0014	2.6	1.6
Die attach ASIC				
Die attach MEMS	300,909	0.0030	56.9	3.6
Wire bond	2,418,061	0.0242	46	28.8
Epoxy lid	573,823	0.0057	11	6.8
Sub-total	5,257,252	0.037	100.0	44
QFN-12 packages	4,706,361	0.0470		56
Total	8,404,408	0.084		100

Table 6.7 Capital and throughput assumptions for cost calculation

Wafer diameter	200 mm	Throughput		Maximum
QFN 12-pin package	$/unit	Wafers/hour	Dies/hour	
Dicing	100,000	25.5		192,780 (wafers)
Die attach ASIC	250,000		9,000	68,040,000
Die attach MEMS	250,000		9,000	68,040,000
Wire bond	450,000		2,250	17,010,000
Epoxy lid	100,000		7,920	59,875,200

overall packaging costs attributed to the individual processes by MEMSCOST is listed in Table 6.6. The capital cost of equipment and throughput assumed for the cost estimate are summarised in Table 6.7.

Figure 6.18 shows the results schematically, where it can be seen that the cost of the plastic package and wire bonding appear to dominate (56% and 29%, respectively). The percentage costs for the same accelerometer, fabricated in a multi-die format, are shown in Fig. 6.19. While the cost of the plastic package remains a substantial fraction of the overall packaging costs (43%), the cost of wire bonding has increased (from 29% to 43%) as a consequence of having to bond the ASIC and the MEMS dies. The bond wires are gold, and the cost advantages of replacing gold wire with copper have been discussed in Section 6.5.

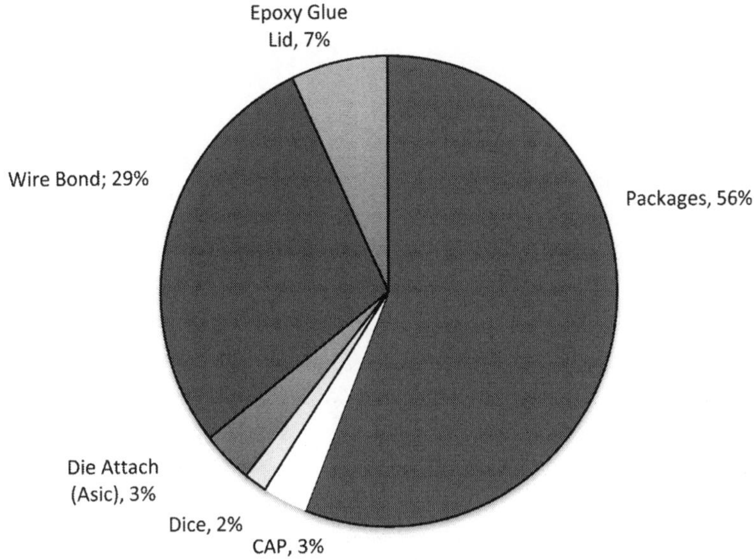

Figure 6.18 Packaging percentages of a single-chip accelerometer.

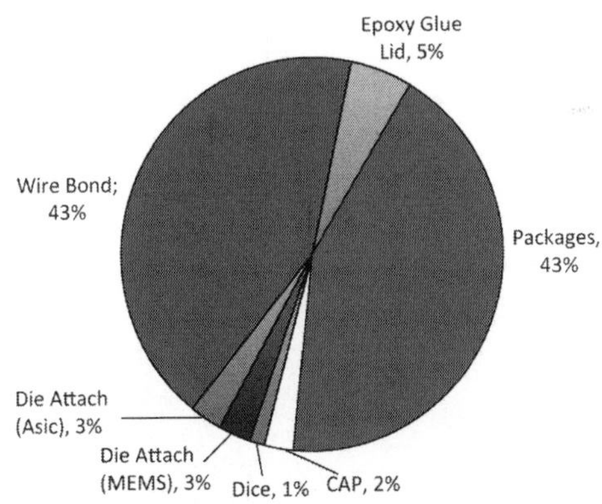

Figure 6.19 Packaging percentages of a multiple-chip accelerometer.

References

Allan, R (2007). Mainstream MEMS: monolithic or hybrid. *Small Times*, January.

Ashby, MF (2010). *Materials Selection in Mechanical Design*. Butterworth Heinemann. ISBN: 978 1856176637, Fig. 4.17.

Dixit, V, et al. (2006). Wire bonding considerations. *Adv. Packaging*, **15**(6), 28.

Esashi, M (2008). Wafer level packaging of MEMS. *J. Micromech. Microeng.*, **18**(7), 1–13.

Evans, D (2004). Advances in MEMS packaging. *Adv. Packaging April*.

Gilleo, K (2005). *MEMS/MOEMS Packaging – Concept, Design, Materials and Processes*. McGraw Hill Nanoscience & Technology Series.

Hsu, T-R (2004). MEMS packaging. *IEE INSPEC*.

Lares, M (2008). Laser dicing cuts wafers from the inside out.www.photonics.com/Article.aspx?AID=31907.

Lewis, F, et al. (2011). Dicing of sensitive MEMS devices: challenges and solutions. *Teledyne Dalsa,* www.sensor mag.com

Lindros, V, et al. (2010). *Handbook of Silicon Based MEMS Materials and Technologies. Micro & Nano Technology Series*. Elsevier. 601.

Maluf, N, and Williams, K (2004). *An Introduction to Microelectromechanical Systems Engineering, 2nd Edition Die Attach Processes*. Artech House Microelectromechanical Library. 225.

Mathew, L (2011). Developments in copper wire bonding. *Int. Microelectron. Packaging Soc.*, May 12.

Panczak, A (2011). Copper wirebonding OSAT viewpoint. *Int. Microelectron. Packaging Soc.*, May 12.

Robinson, P (2012). What is flip die die attach. www.twi.co.uk/services/technical-information/faqs/faq-what-is-flip-die-die-attach.

Chapter 7

Clean Rooms, Buildings and Plant

7.1 Introduction

Integrated circuit (IC) and microelectromechanical systems (MEMS) fabrication must be carried out in specially constructed clean rooms where particles per cubic meter and their sizes in the air are held to a very low level. This is to avoid contamination of the MEMS wafers and devices during fabrication as air-borne particles are generally bigger than the smallest MEMS features and can cause fabrication errors and low yield.

A typical clean-room construction consists of a re-circulating air system, whereby air, entering the clean room, is cleaned by high-quality filters (high-efficiency particulate air filters known as HEPA filters). The air usually flows vertically from the ceiling, down through the perforated floor panels, and is then recycled back through the ceiling.

The US Federal Standard for air quality (FED STD 209E) has been replaced by the International Organization for Standardization (ISO) 14644-1 but is still widely quoted. The two designations are shown in Table 7.1. The estimated cost of clean rooms depends on the class required, as shown in Fig. 7.1 ($\$/ft^2$).

MEMS Cost Analysis: From Laboratory to Industry
Ron Lawes
Copyright © 2014 Pan Stanford Publishing Pte. Ltd.
ISBN 978-981-4411-06-6 (Hardcover), 978-981-4411-07-3 (eBook)
www.panstanford.com

Table 7.1 Clean-room standards for particulates

Class		Max. particles/m^3					
US FED 209E	ISO	≥0.1 μm	≥0.2 μm	≥0.3 μm	≥0.5 μm	≥1 μm	≥5 μm
	ISO 1	10	2				
	ISO 2	100	24	10	4		
Class 1	ISO 3	1,000	237	102	35	8	
Class 10	ISO 4	1×10^4	2,370	1,020	352	83	
Class 100	ISO 5	1×10^5	2.4×10^4	1×10^4	3,520	832	29
Class 1,000	ISO 6	1×10^6	2.4×10^5	1×10^5	3.5×10^4	8,320	293
Class 10,000	ISO 7				3.5×10^5	8.3×10^4	2,930
Class 100,000	ISO 8				3.5×10^6	8.3×10^5	2.9×10^4
General air	ISO 9				3.5×10^7	8.3×10^6	2.9×10^5

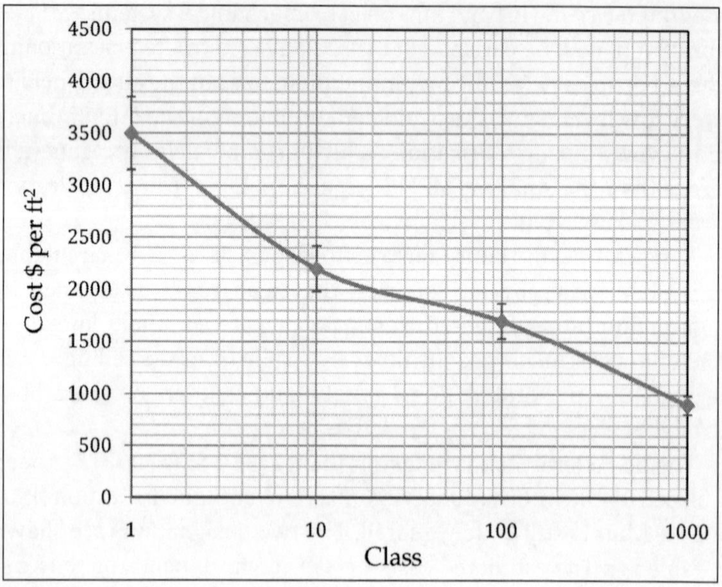

Figure 7.1 Cost of clean rooms as a function of the class of cleanliness ($/ft^2).

The class of clean rooms is a function of the particulate size and is calculated from the formulae

$$P_N = 10^N (0.1/D)^{2.00} \tag{7.1}$$

and

$$P_N = 35.15M(0.5/D)^{2.2} \tag{7.2}$$

where:

P_n = maximum permitted particles per cubic metre of the specified size (rounded to the nearest whole number)
D = size of particulates in microns
N = ISO number (less than 9, multiples of 0.1)
M = class number

From Eqs. 7.1 and 7.2 it follows that, approximately:

$$N = 3 + Log_{10} M \tag{7.3}$$

Figure 7.2 shows a schematic of a MEMS foundry, illustrating the key areas of fabrication, testing and packaging. In this version, the plant (utilities) necessary to run the clean-room air-conditioning and services for the equipment is on the first floor, the clean rooms

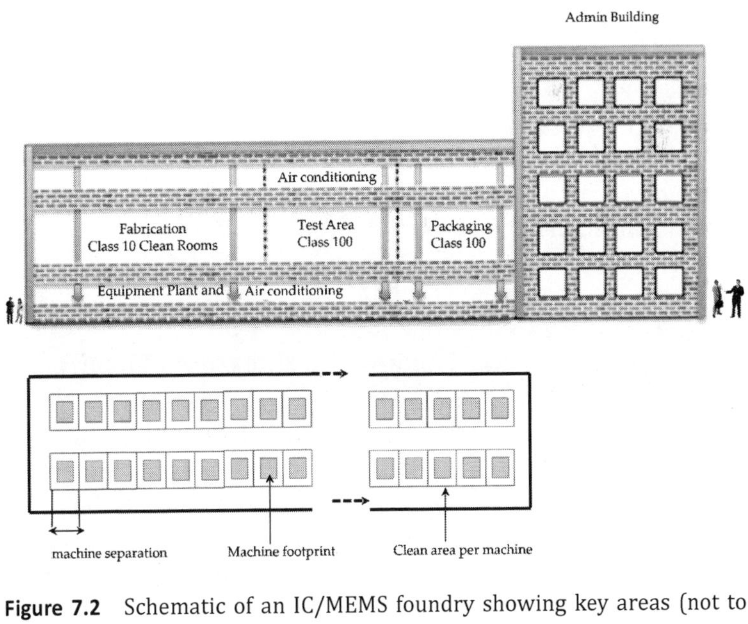

Figure 7.2 Schematic of an IC/MEMS foundry showing key areas (not to scale).

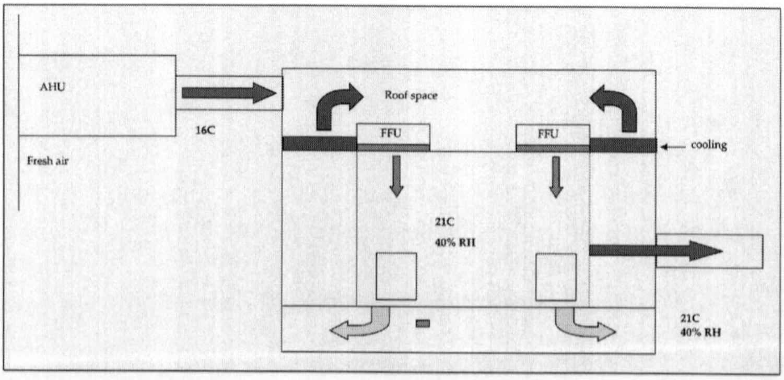

Figure 7.3 Schematic of an IC/MEMS foundry showing basic air-conditioning (courtesy Prof. Stephenson, Scottish Microelectronics Centre).

are on the second floor and the plenum chambers for the air-conditioning on the third floor. This arrangement gives vertical laminar flow air to the clean rooms (see Fig. 7.3). An alternative arrangement consists of two levels, with the plant at the same level as the clean rooms and its equipment but with the services in separate chase areas.

A notional administration building is included in the area and cost estimates, its size calculated from the estimated support staff and the nominal area assigned to each staff member. In addition, the model assumes a boardroom, conference facilities, storage and goods in/goods out space. These costs are not included in the wafer manufacturing costs as they would typically be included in overheads.

7.2 Clean Rooms and Buildings for Integrated Circuits

The semiconductor industry has a wealth of data and studies on clean rooms, buildings and energy consumption. Over the last two decades, new facilities have been built to accommodate both the increased demand for ICs and the rapid technological progress, where more expensive equipment is needed to manufacture new technology nodes and to allow for the increase in wafer diameter 150 mm through 200 mm to 300 mm.

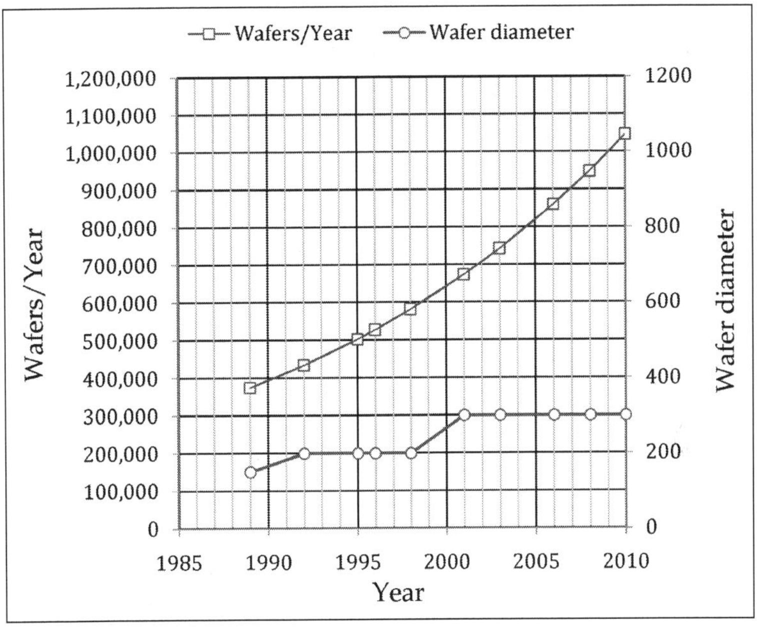

Figure 7.4 Wafers/year and wafer diameter as a function of the technology node.

Cost estimates are sensitive to several assumptions, in particular the designed maximum capacity per year, the die size and the number of mask levels (and hence the number of process steps). The estimated annual output of a typical semiconductor foundry is shown in Fig. 7.4, along with the wafer diameter (150 mm, 200 mm and 300 mm).

MEMSCOST has a Clean Room and Buildings Cost Simulator, which calculates the number of machines required for fabrication, testing and packaging. It bases this calculation on the number of wafers or dies required per year, the estimated die size and the technology node to be introduced to the market. The cost of the clean rooms, the plant and the air-conditioning building is then calculated from the area required, its class of cleanliness and the $/ft^2 assumptions. Also included is the cost of a building to provide accommodation for the direct staff and the administration and management staff and to provide space for goods in and goods out. An example of a MEMSCOST calculation for clean rooms and buildings is shown in Table 7.2 and is based on a 0.18 μm technology node, assuming 500,000 wafers per year, a die area of 100 mm^2 and a yield of 95%.

MEMSCOST can estimate the cost of the building, as specified, and can calculate the capital cost of the equipment to meet the estimated demand. In its present form MEMSCOST assumes that the installed plant for the various utilities costs $260,000 per 100,000 wafers designed capacity. This is derived from available Sematech data (International Sematech, 2002), which estimates the utilities for a 0.18 μm, 200 mm, 360,000 wafers/year factory to cost $92 million (utilities include electrical power, ultra-pure water, cooling water, heating systems, scrubbers for toxic gases, vacuum systems, gases and air-conditioning).

Table 7.3 estimates the total cost of constructing the factory for 500,000 wafers per year in a 200 mm factory, which includes building, purchasing and installing the plant and equipment.

The high level of energy consumption necessary to run the equipment and plant is a significant fraction of the running cost of a semiconductor factory. For example, in addition to basic heating and lighting, it requires energy for the fabrication, test and packaging equipment, for the equipment supplying gases and liquids and for the air-conditioning plant, which has to absorb the heat load of the installed equipment and maintain the temperature and humidity of the clean rooms (220°C +/−10°C, relative humidity 40%). Figure 7.5 shows the approximate percentage of energy used in a semiconductor factory, designed for 200 mm wafers, with a total power consumption of 24.4 MW (214 kWh for three-shift operation (Huang, 2008)).

In the early years of the growth of the semiconductor industry, it was noticed that the cost of building new semiconductor factories followed an exponential relationship (Ross, 2003). This relationship is known as Rock's law and predicts 'the cost of an IC factory will double every 4 years'. Figure 7.6 shows the cost of constructing a state-of-the-art factory in the year that a more advanced technology node was introduced, as estimated by various authors and by MEMSCOST. The estimated cost of the buildings, equipment and plant are shown separately. Earlier estimates in the literature

Table 7.2 Clean Room and Buildings Cost Simulator. The typical cost for clean rooms of varying class and size for a 500,000, 200 mm wafers/year IC facility equipped for a 0.18 μm technology node

Simulator	Class	\$/ft^2	Machines	Length ft	Width ft	Area ft^2	Cost \$ million
Fabrication	1	3500	171	1032	30	30,960	108.36
Testing	100	1700	14	84	30	2,520	4.28
Packaging	100	1700	5	36	30	1,080	1.84
Total clean rooms			190	1152	30	34,560	114.48
Footprint area per machine ft^2				100			
Brickwork shell		250		1172	50	58,600	14.65
Plant area		250		1172	50	58,600	14.65
Air-con.		250		1172	50	58,600	14.65
Total brickwork shell			1172		50	178,800	43.95
Gap between clean-room wall and building wall		10					
Total manufacturing			210,360	188.43			
Direct staff/shift	86						
Admin. building	500						
Total staff		172					
Area/staff ft^2	50						
Office area required ft^2						>8,600	
Floor area (1) ft^2				90	30	4,500	2.25
Floor area (2) ft^2				90	30	4,500	2.25
Boardroom + conference/sales facilities				90	50	4,500	2.25
Materials in + goods out + workshops				90	30	4,500	2.25
No. of floors	4						
Total admin. building						18,000	9.0
Total factory						228,360	167.4

Table 7.3 Cost of a semiconductor factory (see text for assumptions)

Component	Cost
190 machines for fabrication, testing and packaging (MEMSCOST)	$\$584 \times 10^6$
Clean rooms, production and admin buildings (Table 7.2)	$\$167 \times 10^6$
Plant, services (260 × 500,000)	$\$130 \times 10^6$
Total	$\$881 \times 10^6$

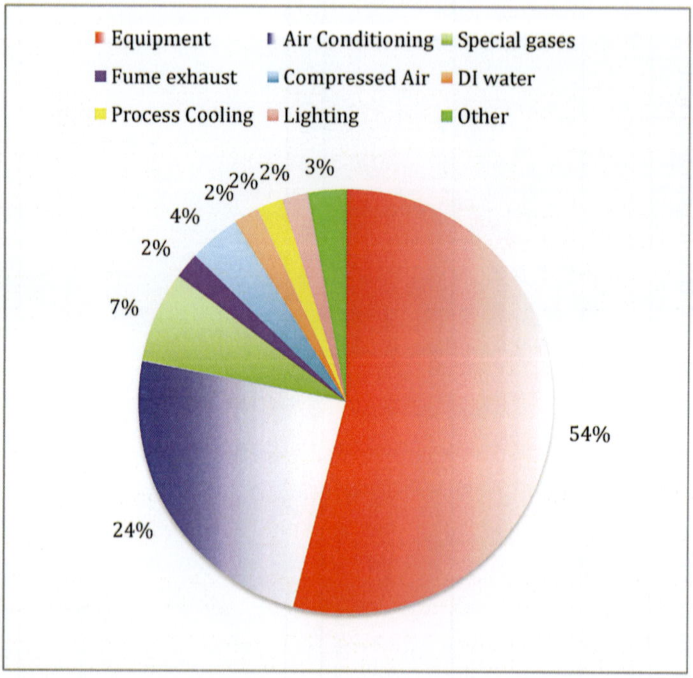

Figure 7.5 % energy usage for a 200 mm semiconductor factory.

are shown—*Communications Engineering Serie* (Huartarte, 2007) and *IC Knowledge* (Jones, 2010), along with some industry data. Considering the significant technology developments and a major jump from 200 mm to 300 mm wafer production, Rock's law has held up well for the last two decades, from 1988 (0.8 μm) to 2010 (0.032 μm).

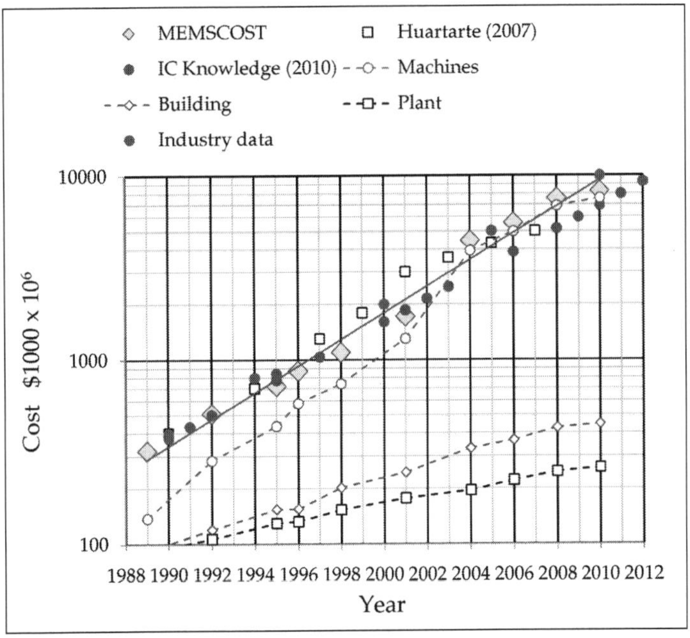

Figure 7.6 Cost of IC factories as a function of the year of the technology node introduction (Rock's law) compared with the industry's data.

MEMSCOST can generate the likely cost of running an IC factory, using assumed power and energy loading for the number of machines required to meet demand. It is assumed that the ratio of energy used by the manufacturing machines to all the services (air-conditioning, etc.) is 40:60 for 150 mm and 200 mm wafers and 50:50 for 300 mm wafers over the period shown, 1989–2010.

As an example, Table 7.4 shows the annual cost for a factory producing 500,000 200 mm diameter wafers per year, equipped for a 0.18 μm technology node. This requires 215 MWh of energy per year. In 2011, the cost of electricity in the United States was $0.072 per kWh (*Electric Power Monthly*, 2011), resulting in an annual electricity cost of approximately $15.5 million. MEMSCOST automatically includes an estimate for the cost of energy in all device cost calculations

Figure 7.7 shows estimates for the energy consumption, in kWh per cm^2, for IC wafer factories over the period 1989–2010. The

Table 7.4 Running costs for a factory producing 500,000, 200 mm diameter wafers per year equipped for a 0.18 μm technology node

Running costs	kW	kWh/year	Annual cost 0.072 $/kWh
Machines			
Fabrication (40%)	12,226	86,167,356	6,204,050
Clean rooms			
Fabrication (60%)	18,189	129,251,034	9,306,075
Total	30,315	215,418,390	15,510,125
kWh/cm^2		1.37	
$/cm^2			0.099

data is approximate and comes from a variety of academic and industrial sources, mainly from the United States, Korea and Taiwan (Hu, 2003; Williams, 2004; Hu, 2008; Czerniak, 2010). A MEMSCOST estimate over the period 1989–2010 is shown for comparison with the targets set by the International Technology Roadmap for Semiconductors (ITRS, 2006) and with recent data from Infinion's Villach factory in Germany (Infinion, 2011).

The energy consumed per wafer has increased with time, averaging 260 kWh per wafer for 150 mm wafers in 1989 to 380–400 kWh per wafer for 200 mm and 460–590 kWh per wafer for 300 mm wafers (MEMSCOST estimates). Over this same period the energy density requirements have reduced from an estimated 1.9 kWh per cm^2 for 150 mm wafers to 1.2-1.4 kWh per cm^2 for 200 mm wafers to 0.6–0.8 kWh per cm^2 for 300 mm wafers. This is despite the need for reduced particle counts to maintain cost-effective yields for reducing technology nodes (180 nm to 65 nm) and is mainly due to the close attention being paid to efficient energy use.

7.3 Clean Rooms and Buildings for a MEMS Foundry

Data is sparse on the costs of constructing a building for a MEMS foundry, purchasing and installing of the capital equipment and the subsequent running cost for wafer production. The total cost will depend on whether the foundry is designed for large-

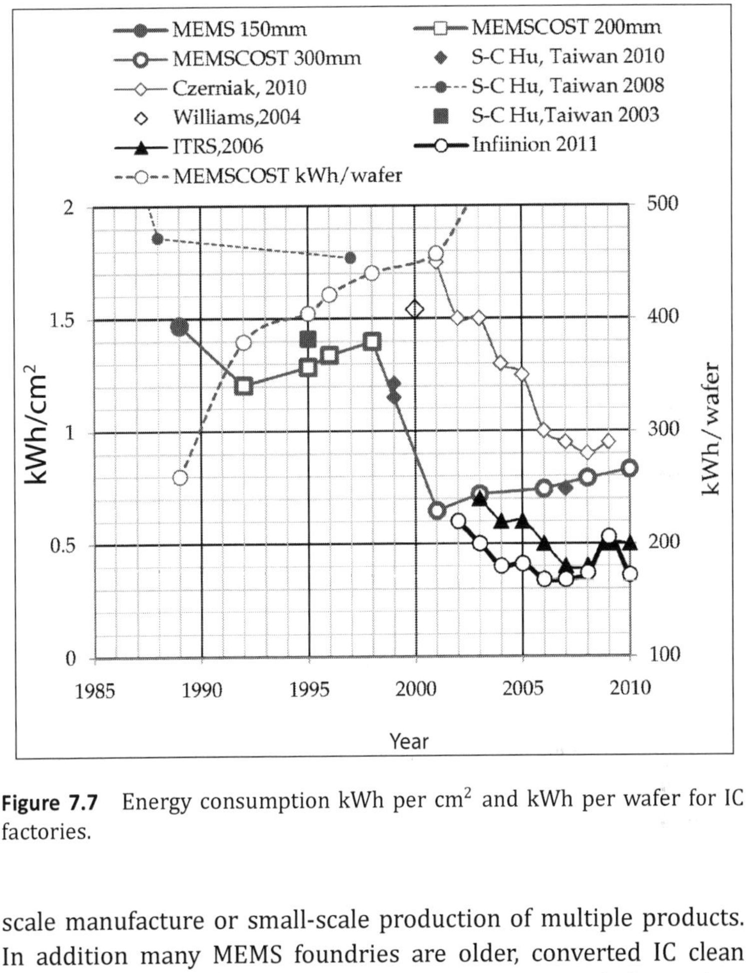

Figure 7.7 Energy consumption kWh per cm^2 and kWh per wafer for IC factories.

scale manufacture or small-scale production of multiple products. In addition many MEMS foundries are older, converted IC clean factories or previously had not been constructed to such demanding energy efficiency specifications. New purpose-built factories will be more energy efficient.

In the absence of MEMS-related data, it is assumed that the basic parameters that have been calculated for semiconductor factories (Section 7.2) can be scaled and applied to a MEMS foundry. Table 7.5 calculates the area and cost of buildings for a 200 mm wafer diameter MEMS foundry designed to produce 200 million

Table 7.5 Clean Room and Buildings Cost Simulator. The typical cost for clean rooms of varying class and size for a 10,000 200 mm wafers/year IC facility equipped for a 0.18 μm technology node

Simulator	Class	$/ft^2	Machines	Length ft	Width ft	Area ft^2	Cost $ million
Fabrication	1	3500	20	120	30	3,600	12.6
Testing	100	1700	10	60	30	1.800	3.06
Packaging	100	1700	5	36	30	1,080	1.84
Total clean rooms		35	216	30	34,560	17.5	
Footprint area per machine ft^2				100			
Brickwork shell							
Clean rooms		250		236	50	11,800	2.95
Plant area		250		236	50	11,800	2.95
Air-con.		250		236	50	11,800	2.95
Total brickwork shell				236	50	35,400	8.85
Gap between clean-room wall and building wall		10					
Total manufacturing						41,880	26.35
Direct staff/shift	10						
Admin. building	500						
Total staff		20					
Area/staff ft^2	50						
Office area required ft^2						>1,000	
Floor area (1) ft^2				90	50	4,500	2.25
Floor area (2) ft^2				90	50	4,500	2.25
Boardroom + conference/sales facilities			50	50		4,500	2.25
Materials in + goods out + workshops			50	30		4,500	2.25
No. of floors	4						
Total admin. building				Area		18,000	9.0
Total factory						59,880	35.4

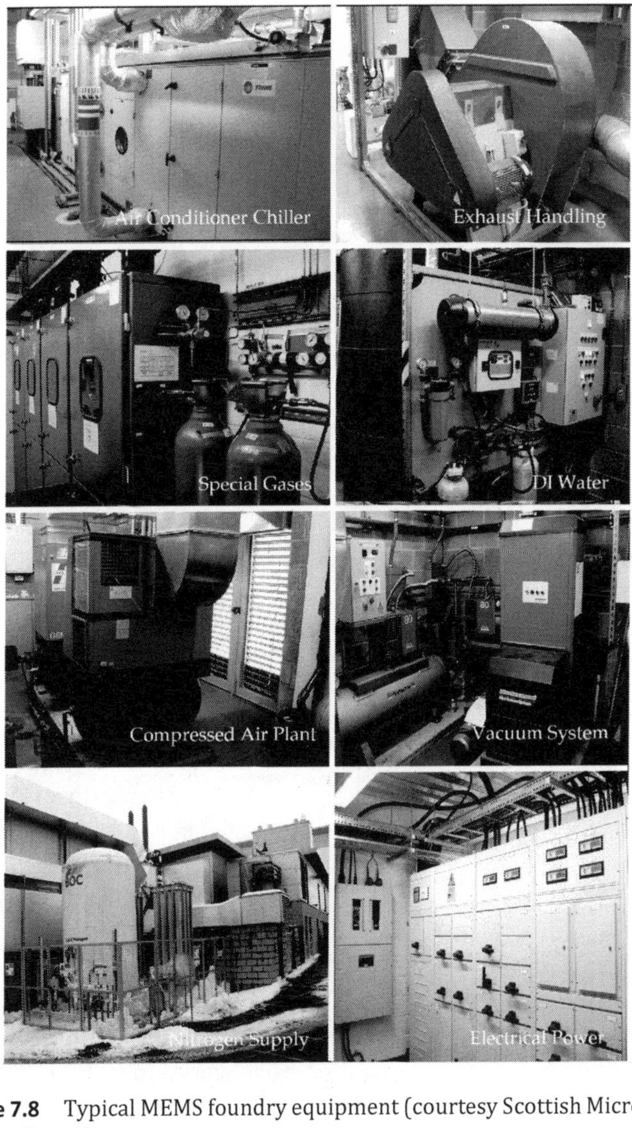

Figure 7.8 Typical MEMS foundry equipment (courtesy Scottish Microelectronics Centre).

Table 7.6 Cost of a MEMS factory (see text for assumptions)

Component	Cost
Machines for fabrication, testing and packaging (MEMSCOST)	$\$34.9 \times 10^6$
Clean rooms, production and admin. buildings (Table 7.5)	$\$35.4 \times 10^6$
Plant, services ($10,000,000 + 260 × 10000)	$\$12.6 \times 10^6$
Total	$\$82.9 \times 10^6$

2 mm × 2 mm single-die accelerometers per year from 10,000 wafers at 85% yield.

Typical plant equipment for a MEMS factory/foundry is shown in Fig. 7.8.

Using MEMSCOST, a similar calculation for constructing a MEMS factory can be made from the data calculated in Table 7.5, that is, a factory producing MEMS device/devices complete with 0.18 μm technology for signal-processing dies. The number of wafers per year produced by a MEMS factory is much lower than for an IC factory, so the cost of the plant (C_{PL}) to supply the utilities must be modified for low throughput (N wafers per year). MEMSCOST assumes that the minimum plant would cost $10 million with a rate of $260 per wafer thereafter, that is,

$$C_{PL} = 10,000,000 + 260N \tag{7.4}$$

The total cost of buildings, equipment and plant for a 10,000 wafers per year on a 200 mm diameter wafer MEMS foundry is shown in Table 7.5. In this example, the investment required to set up a MEMS factory is an order of magnitude less than that required for an IC factory ($82.9 million compared with $881 million).

The cost of setting up a MEMS factory (both application-specific integrated circuits (ASICs) and MEMS components) has not followed the rapid increase shown by the IC industry, possibly due to the modest number of wafers required each year to meet demand, and hence the reduced size of clean rooms and plant needed to support the equipment. As an example, Fig. 7.9 shows the increasing cost of constructing a MEMS factory to meet increasing demand.

A different cost structure emerges if a MEMS factory concentrates on the MEMS component, leaving specialist foundries to manufacture the signal-processing ASICs. Figure 7.10 shows the cost of

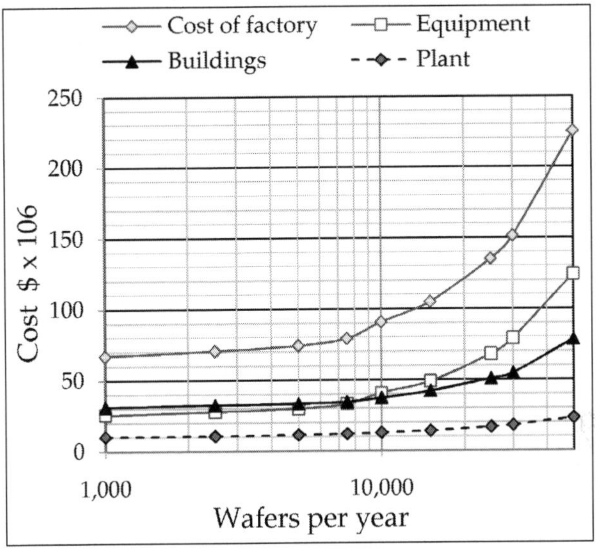

Figure 7.9 Cost of constructing a 200 mm MEMS foundry.

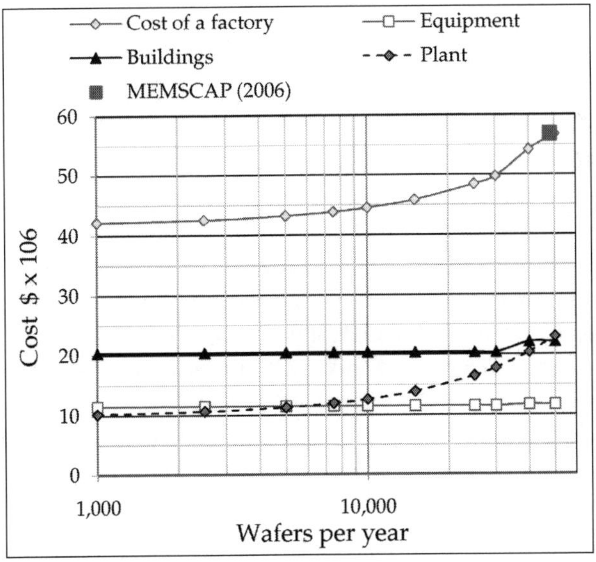

Figure 7.10 Cost of constructing a 200 mm MEMS foundry (MEMS components only).

Table 7.7 Energy costs for a factory producing 2 mm × 2 mm dies on 150 mm diameter wafers

Running costs	kW MEMSCOST	kWh/year MEMSCOST	Cost $/year 0.072 $/kWh	kW ADI
Machines				
Fabrication (40%)	581	2,692,935	193,891	612*
Clean rooms				
Fabrication (60%)	872	4,039,403	290,837	845*
Standby	80	236,000	16,992	
Total	1,533	6,966,338	501,720	1,457*
kWh/cm^2		1.31		1.53*
kWh/wafer		232		271

*See Branham & Gutowski, 2010.
Abbreviation: ADI, Analog Devices Inc.

constructing such a factory/foundry as a function of the designed capacity (2011 costs). As the throughput increases, the equipment and buildings needed remain unchanged but the plant required to supply the utilities increases. Note the example of the cost of a MEMSCAP foundry with a designed capacity of 48,000 wafers per year (Semiconductor Technology, 2006).

Energy costs have been published by Analog Devices, micromachined products division (Branham & Gutowski, 2010), which enables a comparison to be made with a MEMSCOST calculation. The factory contains a 150 mm wafer diameter wafer line that is assumed to produce 2 mm × 2 mm single die MEMS devices on 30,000 wafers per year. The data shown in Table 7.7 summarises the electrical power consumed and is based on the author's averaging of the data. Assuming 30,000 wafers per year, MEMSCOST estimates 4,635 hours of production and 2,925 hours of standby. The energy consumption is estimated by MEMSCOST at 1.31 kWh per cm^2 and is in reasonable agreement with the published value of 1.53 kWh per cm^2.

MEMSCOST can be used to estimate the energy consumption per wafer size. Figure 7.11 illustrates the estimated energy consumption in kWh per wafer for 150 mm and 200 mm wafers as a function of throughput in wafers per year. The average for 150 mm is 223 kWh per wafer and for 200 mm 473 kWh per wafer, corresponding to 1.26

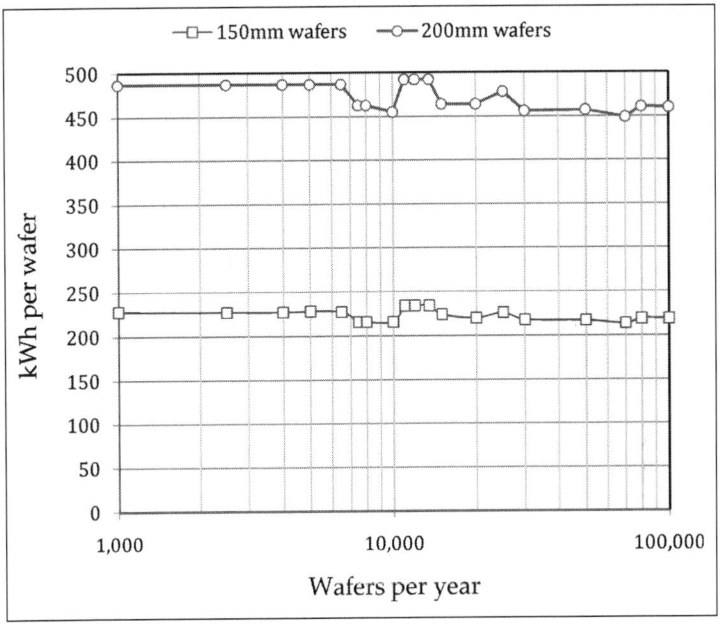

Figure 7.11 Energy consumption per wafer for a MEMS factory.

kWh per cm^2 and 1.5 kWh per wafer. The perturbations in the data arise as extra machines are installed to meet the demand for more output.

References

Branham, M, and Gutowski, T (2010). Deconstructing energy use in microelectronics manufacturing: an experimental case study of an MEMS fabrication facility. *Environ. Sci. Technol.*, **44**(11), 4295–4301.

Czerniak, M (2010). Vacuum/abatement technology saves the bottom line and the planet. *Solid State Technol.*, **53**(9).

(2011). *Electric Power Monthly*, June, Table 4.

Huartarte, J, et al. (2007). *Understanding Fabless IC Technology. Communications Engineering Series.* Table 1.14, 22.

Hu, S-C, et al. (2010). Characterization of energy use in 300 mm DRAM (Dynamic Random Access Memory) wafer fabrication plants (fabs) in Taiwan. *Energy*, **35**, 3788–3792.

Hu, S-C, et al. (2008). Power consumption benchmark for a semiconductor cleanroom facility system. *Energy Buildings*, **40**, 1765–1770.

Huang, T (2008). Strategies for energy reduction in semiconductor manufacturing. *ElectroIQ J.*

Hu, S-C (2003). Power consumption of semiconductor fabs in Taiwan. *Energy*, **28**, 895–907.

International Sematech (2002). Fab utility cost values for cost of ownership (COO) calculations. *International Sematech Technology Transfer 02034260A-TR.*

ITRS (2006). ITRS roadmap for semiconductors 2006 update: environment, safety and health. http://www.itrs.com

Infinion (2011). Updated enviromental statement 2011 of the Villach site. Infinion Technologies Austria AG.

Jones, SW (2008). Exponential trends in the integrated circuit industry. *IC Knowledge.* www.icknowledge.com/trends/exponential3.pdf.

McGregor, J (2007). The common platform technology: a new model for semiconductor manufacturing. *In-Stat In-Depth Anal.*

Nenni, D (2011). Semiconductor fab Wiki. Semiconductor WikiI project.

Ross, PE (2003). The rules engineers live by weren't always set in stone. *IEEE Spectrum December.*

Schrecengost, R, and Naughton, P (2004). Cleanroom energy optimization methods. *Proc. 14th Symp. Improving Building Syst., Hot Humid Climates.*

Semiconductor Technology (2006). MEMSCAP MEMS wafer fab, Bernin, France.www.semiconductor_technology.com/projects/memscap

Williams, E (2004). Energy intensity of computer manufacturing. *Environ. Sci.*, **38**, 6166–6174.

Chapter 8

The MEMSCOST Spreadsheet

8.1 Introduction

Cost analysis calculations can be readily undertaken using a conventional spreadsheet package, such as Microsoft Excel. MEMSCOST is one such spreadsheet, developed by the author and used for all the calculations in this book. MEMSCOST originates from the author's efforts in the 1990s to establish the cost of microelectromechanical systems (MEMS) fabrication at the research and development (R&D) stage. Within the last five years, the capability of MEMSCOST has been much expanded. It includes a family of high-aspect-ratio (HAR) processes and other MEMS fabrication techniques particularly relevant to the industry, along with device testing and packaging techniques.

MEMSCOST analyses the cost of MEMS manufacturing in three areas:

(i) **FABRICATION** of the device using semiconductor and HAR equipment and processes
(ii) **TESTING** of the device at various stages of manufacture
(iii) **PACKAGING** of the device in either a plastic or a ceramic standard package

MEMS Cost Analysis: From Laboratory to Industry
Ron Lawes
Copyright © 2014 Pan Stanford Publishing Pte. Ltd.
ISBN 978-981-4411-06-6 (Hardcover), 978-981-4411-07-3 (eBook)
www.panstanford.com

The relevant equipment parameters are the capital cost of the equipment and its throughput in wafers/hour. From this data, the manufacturing costs can be calculated on a 'fixed plus variable' basis (see Section 8.11). Fabrication equipment costs for MEMS are reasonably well known. Those for testing and packaging are less visible but are analogous to the semiconductor industry.

It is difficult to include data that would be applicable to all equipment and processes used in the MEMS industry. Processes for a given product are not widely publicised, and a considered view, based on an understanding of MEMS technology, is involved. The capital cost of equipment is not widely publicised, and the type of equipment will vary between factories.

High-volume production with wafer-track automation is more applicable to large in-house facilities, while single-station equipment may be more applicable to foundries and start-ups. In addition, a foundry or start-up can take advantage of the wealth of high-quality, low-price, second-hand equipment coming out of obsolescent semiconductor factories. The introduction of the MEMSCOST machine simulators goes some way to providing reasonable default data on throughput (wafers per hour) but makes provision for user-defined data to be substituted.

The initial design of a MEMS device will depend mainly on the engineering requirements to produce working devices and the materials needed for their construction. The ability to manufacture the design will depend on the process technology and the equipment available in an in-house or commercial foundry. There may be various options, and cost analysis can be used to understand, and perhaps reduce, the cost of fabrication, testing and packaging.

8.2 Basic Structure of the MEMSCOST Spreadsheet

The key components of MEMSCOST are shown in Fig. 8.1 and include

- *Process Libraries* to describe all the steps required during manufacture;

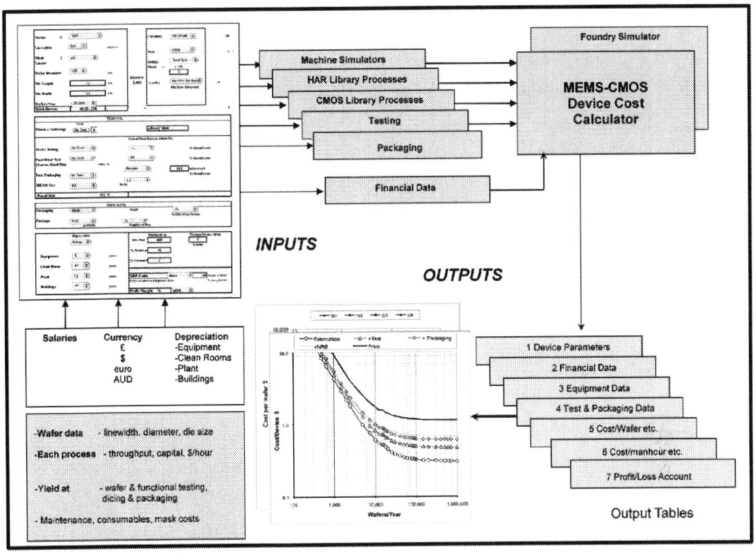

Figure 8.1 Schematic diagram for the basic structure of MEMSCOST.

- *Machine Simulators* to determine wafer throughput for a given thickness of material to be etched, deposited or otherwise processed;
- a *Device Cost Calculator* that uses capital costs and throughput data for the fixed-plus-variable cost algorithms (as outlined in Section 8.11) to estimate various costs;
- *Output Tables* providing costing and overview data for management purposes; and
- a *Master Set-Up* list that drives MEMSCOST at the top level and enables basic data to be varied (e.g., wafer diameter) and reports basic costs (e.g., unit cost per device).

8.3 Master Set-Up

MEMSCOST is driven by the user through the *Master Set-Up* facility, which enables the main features of a device to be specified, for example, wafer and die size, wafer output per year and testing and packaging regimes. Basic financial data such as depreciation assumptions, hours worked per operator/technician and the salary and overheads associated with each production staff member is included. Table 8.1 shows the list of options.

The details of each device to be costed can be added to the *Process Library* and subsequently recalled by referencing the device name. The wafer diameter, die dimensions and mask levels can be set, which, along with the proposed number of wafers per year and expected yield, will enable the number of good dies to be calculated from the number of die sites per wafer.

The testing regime can be constructed from four possibilities, namely, a visual/metrology test, a wafer process test (via test dies), a functional test of all dies and, lastly, a post-packaging test for either an associated application-specific integrated circuit (ASIC) or the MEMS component or both. Each option can be deleted from the cost estimate, if required. The packaging options are outlined in Chapter 6. They can be defined by the *Process Library*, deleted from the cost estimate or set as a fixed proportion of the fabrication cost.

MEMSCOST assumes that depreciation is linear over a set number of years. It is calculated separately for equipment, clean rooms, plant and buildings at a rate of 5, 10, 15 and 25 years, respectively. These assumptions can be changed, if required.

The cost of manpower per hour is calculated from the hours per year, the average salary level (base salary plus benefits) of the production staff (e.g., operators, technicians, maintenance engineers) and the ratio of machines to production staff.

MEMSCOST offers a selection of 'standard' processes in the *Process Library*, such as those for manufacturing a CMOS signal processor, accelerometer and microphones and a capability to add new processes to the library. Each piece of equipment is identified in the *Process List* and the number of times it is used in the process. The capital cost and throughput of the equipment are derived from the *Capital Cost Library* and the *Machine Simulators*, respectively, and the costs submitted to the *Device Calculator*. The material and clean-room costs, which depend on the equipment footprint and other financial data, are then added to estimate the overall costs.

The data flow is shown in Fig. 8.2. Ideally, the *Process Library*, the *Capital Cost Library* and the *Machine Simulators* should be constructed with specific data for a given product manufactured in

Table 8.1 Options set by the *Master Set-Up* page

Function	Options
FABRICATION	
Device	CMOS, accelerometer, microphone MEMS2 SUMMIT, MUMPS, packaging, etc.
Linewidth (μm)	2, 1. 2, 0.8, 0.5, 0.35, 0.25, 0.18, 013, 0.09, 0.065, 0.045, 0.032 or set by device library
Mask levels	1–40 or set by device library
Wafer diameter (mm)	100, 150, 200, 300
Die length (mm)	Custom
Die width (mm)	Custom
Wafers/year	Custom
Known good die	Calculated from test data
TESTING	
Metrology	Dimensional measurements
Wafer test (%)	On-wafer basic tests
Functional test (%)	ASIC function test (digital)
Functional test (%)	MEMS test
Post-packaging ASIC (%)	ASIC function test (digital)
Post-packaging MEMS (%)	MEMS test
Yield (%)	Pass rate for each test, each process and overall
PACKAGING	
Packaging mode	None, model, %
Package type	QFN, LGA
No. of pins	Selected for available package
DEPRECIATION	
Equipment (years)	None, 3–10
Clean rooms (years)	None, 3–25
Plant (years)	None, 3–25
Buildings (years)	None, 3–25
SHIFT	
Mode	Hours worked or per shift
Weeks/year	50
Days/week	6
Shifts/day	2
Hours/shift	10
Hours/operator	1,600
Uptime (%)	90
SALARY ($/year)	47,000
Overheads (%)	150
Machines/operator	2

Abbreviations: CMOS, complementary-metal-oxide semiconductor; SUMMIT, Sandia ultra-planar, multi-level MEMS technology; MUMPS, multi-user MEMS processes; QFN, quad flat no.; LGA, land grid array.

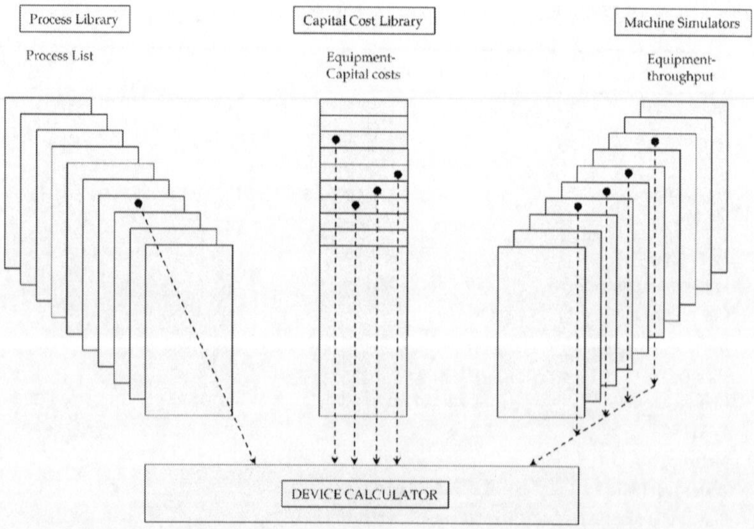

Figure 8.2 Basic data flow from the *Process Library*, the *Capital Cost Library* and the *Machine Simulators*.

a given factory/foundry. When this information is not available the library default entries are used.

8.4 Process Libraries

The MEMSCOST *Process Library* defines separate process lists for fabrication, testing and packaging.

Currently, 13 processes have been defined (there is space for 20 processes), each comprising up to 32 different fabrication machines/processes, 6 test machines/processes and up to 12 packaging/process machines (a total of 50 process options). For each process, the number of individual steps and the associated equipment are defined. For costing purposes no account is taken of the order in which the steps are listed. A notional example of a library entry for a MEMS comb sensor fabrication process is shown in Table 8.2. This process corresponds to that shown in Chapter 4, Fig. 4.5 and the device shown in Chapter 9, Fig. 9.9.

Table 8.2 Typical fabrication library entry for a MEMS comb sensor, including the cap

Fabrication		
	Linewidth	1
	Mask layers	4
HAR processes	Layers	Thickness
X-LIGA-P	–	–
X-LIGA-S	–	–
UV-LIGA–S	–	–
Laser	–	–
DRIE	1	50
Surface microeng. thick poly	–	–
BULK	–	
Wafer bond	1	–
Support equipment		
Resist processes (thin film)	4	0.5
Ion implant		
CMP	1	0.5
Cleaning	3	
Plasma ash	3	0.5
Electrodeposition	–	–
RTA		
Resist processes (thick film)	–	–
Support equipment		
Litho aligner		
Litho stepper	4	0.5
LPCVD poly	1	5
LPCVD silicon nitride	1	0.8
DC sputter metal	1	0.6
LPCVD SiO_2	1	51
Dry etch		
Wet etch (oxide)	4	0.5
Support equipment		
EBL	–	–
Thermal oxide	1	0.6
Passivation	–	–
HF release	1	5
Super-critical-dry	1	
TBD1	–	–
TBD2	–	–
TBD3	–	–
Total process steps	28	

Abbreviations: LIGA, Lithographie, Galvanoformung and Abformung; UV, ultraviolet; DRIE, deep reactive-ion etching; CMP, chemical mechanical polishing; RTA, rapid thermal anneal; LPCVD, low-pressure chemical vapour deposition; EBL, electron-beam lithography; HF, hydrofluoric acid; TBD, to be defined.

Table 8.3 Typical test library entry for a MEMS comb sensor

Test		Yield %
Visual + metrology	0	0
Wafer test (test sites)	1	98
Functional test (ASIC)	1	95
Functional test (MEMS)	1	85
Post-packaging test (ASIC)	0	0
Post-packaging test (MEMS)	1	95
Overall pass		75

For each piece of equipment used in wafer fabrication, the number of steps is defined along with the average thickness of material processed, that is, the thickness of material etched or deposited. The lithography machines (e.g., an aligner, a stepper or both) can be defined for each process, along with the number and cost of masks.

For testing, the library consists of the six different tests available: a visual test, including metrology samples; a test checking the wafer process test sites; an ASIC and/or MEMS functional test of all dies on a wafer; and two post-packaging tests (for the MEMS component of a device and the ASIC component if present). For each test the *Master Set-Up* will define the number of times a given test is used (usually only once) and the associated yield (Table 8.3). The test selection procedure will vary the amount of packaging that has to be done, depending on the pre-packaging yield, or whether the test is done at all (see Chapters 9 and 10 for examples).

For packaging, the library defines the number of times a packaging step is used. When a single die is to be packaged then the die attach process would be used once. For multiple dies the process would be used as many times as necessary. A packaging entry for a single-die device is shown in Table 8.4. The library will contain a large amount of data (e.g., 1,000 variables), so to avoid corruption of data while making minor changes, MEMSCOST has a secure *Entry Library* procedure.

At the top level of security, the *Entry Library* has to be unlocked manually for data entry and manually locked after use. When open,

Table 8.4 Typical packaging library entry for a MEMS comb sensor

Packaging		Yield %
Manufacture special part	–	–
Add CAP @ wafer level	1	99
Die	1	99
Die attach (MEMS)	–	–
Die attach (ASIC)	1	99
Wire bond	1	99
Fill cavity	–	–
Epoxy glue lid	1	100
CVD vapour coat	–	–
TDB1	–	–
TBD2	–	–
TBD3	–	–
Overall yield		96

Abbreviations: CVD, chemical vapour deposition; TBD, to be defined.

the library would normally be in a 'no changes permitted' mode. A selection must then be made defining the type of data modification required. A 'Select All' mode would normally be used to define a complete fabrication, testing and packaging process. Data selection must be made for all 50 cells (fabrication 32, test 6 and packaging 12). These data entries, now in the *Entry Library,* automatically populate the *Process Library*. Once the *Process Library* has been populated, only individual changes may be required. For maximum security of all data, a 'Select Changes' mode is available, where the only permitted access is to the selected individual process steps. Where major changes are required, the 'Select All' mode may again be utilised.

8.5 Fabrication Machine Simulators

Cost calculations for the devices being analysed require input data both for the capital cost and for the throughput of each machine specified in the *Process Library.* Each piece of equipment has a *Machine Simulator* module, which describes the operation of the machine and enables the throughput (wafers layers/hour, thickness of a given material to be deposited or etched) and the maximum wafer layers/year to be set. For example, a wafer stepper's throughput can be set for a fixed value or modelled to take into account the lens field size and magnification, the number of dies/reticles, the stage acceleration and velocity and the exposure time per die. For DRIE, the etch rate can be set or calculated from the aspect ratio and loading factor (ratio of etch to non-etched features) of the microstructures. For wet etching the rates can be set according to the etchant used, such as potassium hydroxide (KOH), ethylene diamine pyrocatechol (EDP) or tetramethyl ammonium hydroxide (TMAH) (see Chapter 4, Table 4.6, for etchant details). Table 8.5 shows the default rates used by MEMSCOST.

8.6 Test and Packaging Machine Simulators

Six different tests may be selected through MEMSCOST, set up singly, sequentially or not at all. MEMSCOST calculates the overall yield at each test step so that the correct wafer and die numbers are available for accurate cost calculation

The tests (in order) are:

- a basic visual (via a microscope) and metrology measurement.
- a wafer test where the properties of a few test sites are measured to ensure wafer fabrication fidelity.
- functional tests for the ASIC and/or MEMS dies, where all dies on a wafer are checked for the designed basic functions. Only marked dies passing all the preceding checks are sent forward for packaging. Testing of any ASIC signal processor can be included here.
- post-packaging functional tests for ASIC and/or MEMS components to test the mechanical and overall integrity of the device after packaging and before shipping.

Any or all of these tests can be selected by the *Master Set-Up* window to be included in the overall test procedure. If selected, MEMSCOST has data for yield, throughput (wafers or dies tested per hour) and materials.

Table 8.5 Equipment/process default rates used in MEMSCOST

Equipment	Rate	
X-ray LIGA SU-8	67	μm/minute
UV-LIGA	15.9	μm/minute
Excimer laser	1.7	μm/min
DRIE	10	μm/minute
Surface micromachining	50	nm/minute
Wet bulk micromachining	800	nm/minute
Wafer bonding	–	
Resist processes (thin film)	60	minutes
Ion implant (batches/hour)	10	minutes
CMP	200	nm/minute
Cleaning	10	minutes
Electroplating	207	nm/minute
RTA	2	minutes
Resist processes (thick film)	60	minutes
Litho aligner	40	wafers/hour
Litho stepper (dies/second)	42	wafers/hour
LPCVD poly	15	nm/minute
Silicon nitride etch	10	nm/minute
DC sputter metal	172	nm/minute
Dry etch	100	nm/minute
Wet etch	500	nm/minute
EBL	0.43	wafers/hour
Thermal oxide	1.4	nm/minute
Passivation	1,200	nm/minute
HF release	60	nm/minute
Critical-point-dry	1	wafer/hour

This data can be entered from three sources:

- The process selected from the *Process Library*.
- A model process based on yield models. At present MEMSCOST only offers standard semiconductor-based formulae for this test.
- A numerical value set by the function keys in the *Master Set-Up* window, which will override any of the above settings.

To date, little research has been reported to establish if any of the standard yield models, developed for semiconductor processes,

Table 8.6 Yield as a function of area x defect density

Name	Formula	Basis
Exponential	$1/1(1+A_\mathrm{D}d)$	Clumping of defects
Poisson	$e^{-A_\mathrm{D}d}$	Random defect distribution
Murphy	$(1-e^{-A_\mathrm{D}d}/A_\mathrm{D}d)^2$	Non-uniform distribution
Seeds	$e^{-\sqrt{A_\mathrm{D}d}}$	
Rectangular	$(1-e^{-2A_\mathrm{D}d})/(2A_\mathrm{D}d)$	Rectangular defect distribution

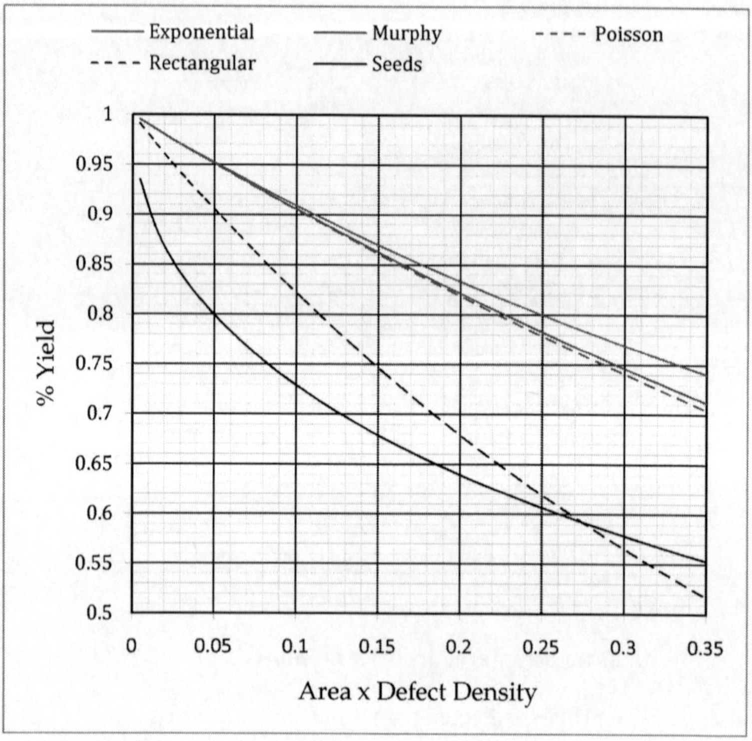

Figure 8.3 Yield as a function of area x defect density.

are relevant to MEMS (see Table 8.6). Figure 8.3 shows the spread of predictions for an integrated circuit (IC) as a function of the product of die area (A_D) and defect density (d) at a particular size defect. Note that the Poisson yield model has been shown to be a good approximation for the accelerometers manufactured at

Analog Devices (Grosjean, 2007). For example, a 1 μm × 1 μm die, manufactured with a defect level of 0.1 mm^2, would have a yield of 90%, predicted by the Poisson model. However, more measured yield data is needed to enable MEMS-specific yield models to be developed. It will be assumed that using the models with modified defect density parameters (defects/cm^2) is a reasonable fit for MEMS devices.

The test modules may be combined in different sequences to evaluate the cost-effectiveness of different strategies. For example, Fig. 8.4 demonstrates two strategies for a notional manufacturing line. At each stage the number of wafers and the pass rate are shown.

For the first option, pre-packaging testing is done on the ASIC and MEMS dies at the wafer level, followed by a final device test on the individual packaged dies. For the second option, no pre-packaging testing is done and only final device testing is undertaken. In both cases 100% device testing is assumed.

The example shown in Fig. 8.4 assumes an annual fabrication of 10,000 wafers, employing a pre- and post-packaging test strategy where the ASIC has a 95% probability of working correctly and the completed device has an 85% probability of passing a wafer-level test. Thus, on average, 8,075 wafers will be marked for packaging (this is a best-case estimate as the faulty dies are likely to be randomly distributed amongst the 10,000 wafers). After the final test of the packaged devices, with a pass rate of 95%, 7,671 of the original 10,000 wafers will have produced working packaged devices, an overall yield of 76.7%. Consider the same output and pass rates where only a post-packaging test strategy is employed. In this case all 10,000 wafers would be sent for packaging and for the final test. On average the same 76.7% of packaged devices will pass the final test.

The economics of the different strategies will depend on the relative cost of each test (i.e., capital cost of the testers and their throughput) and, particularly, the process yield for the ASIC and MEMS dies. For the example shown, significantly different numbers of working dies are packaged in the two options, but the higher costs for packaging may be balanced by the lower cost of testing (e.g., no pre-packaging tests).

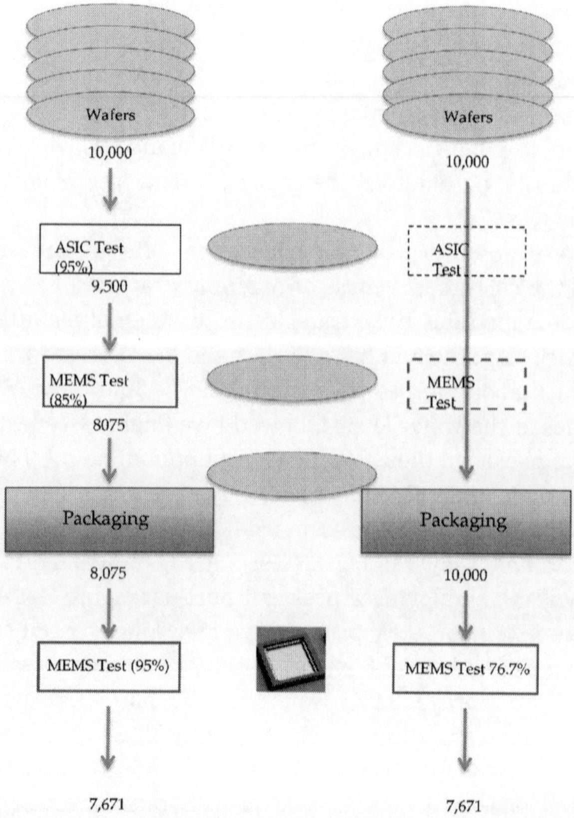

Figure 8.4 Testing strategies. Pre- and post-packaging tests (left side) and post-packaging only (right side).

There is a manufacturing yield above which post-packaging testing alone is cheaper than multiple testing as described by Eqs. 8.1 and 8.2. Assuming $C_W = C_T$ then:

$$nC_T + \alpha nC_P + \alpha nC_T = nC_P + nC_T, \tag{8.1}$$

which reduces to

$$\alpha = C_P/(C_P + C_T) \tag{8.2}$$

where:

n = number of dies per year
C_W = cost of test at wafer level per die

C_T = cost of test at device (packaged) level per die
C_P = packaging cost per die
α = yield

Verification of this approximation is given in Chapter 9 for accelerometers and Chapter 10 for microphones.

8.7 CMOS/BiCMOS Simulation

For devices such as accelerometers, gyros and microphones, any cost estimate must include not only the MEMS component but also the ASIC signal processor. MEMSCOST has a CMOS/BiCMOS simulator to estimate the process steps that are required for an ASIC signal processor manufactured with different technology nodes.

However, accurate simulation of the cost of the various technologies and comparison with market price is not only beyond the scope of this publication but also is beset by price fluctuations from year to year. The introduction of new technology nodes every few years and the general lack of information on manufacturing costs in the public domain make accurate cost estimates difficult. There is data available from commercial organisations specialising in analysing process steps for a wide range of industrial fabrication processes (IC Knowledge, 2012), but such data is not available to MEMSCOST.

A solution for MEMSCOST has been to produce a process list for a simple model (0.8 um CMOS), then to add updated steps, where known (e.g., number of mask, implantation and metallisation steps) and finally to scale the remainder of the process parameters so that the total steps match those estimated for a given technology node. The basis of the model is shown in Fig. 8.5.

A core BiCMOS/CMOS process has been analysed in detail and a process list drawn up to contain the number of times a particular fabrication step is used and hence the total number of fabrication steps (e.g., 0.8 μm, 12 masks, 100 steps). The model is normalised to produce approximately 200 steps for 0.18 μm technology.

Additional data is available for certain major processes developed to advance to the next technology node. Figure 8.6 shows the number of masks/reticles per set for a given technology node as

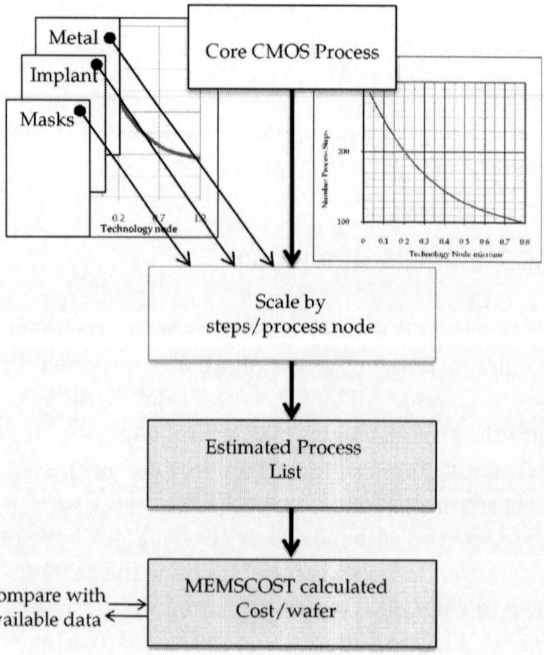

Figure 8.5 MEMSCOST model for the number of steps per technology node.

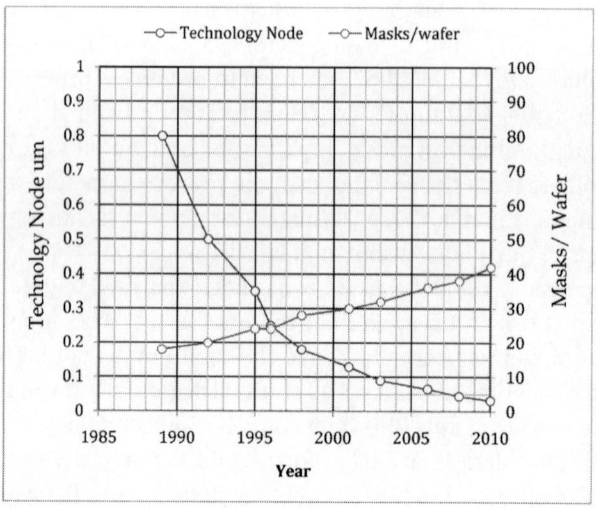

Figure 8.6 Relationship between the technology node, the year of its introduction and the number of masks/wafer.

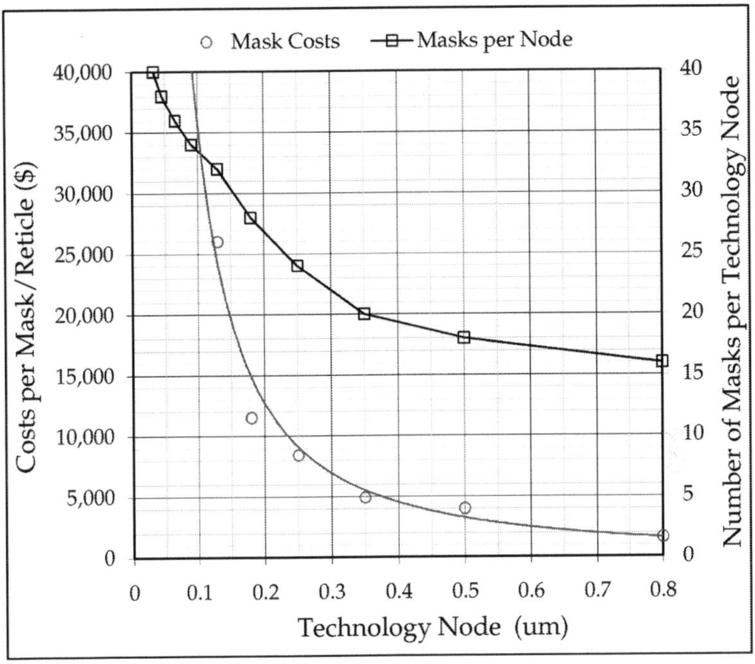

Figure 8.7 Cost of masks/reticles and the number of masks/reticles per technology node.

a function of the year. The cost of masks/reticles and the number of masks/reticles, as a function of the technology node, are shown in Fig. 8.7. The number of implantation steps and metal layers as a function of the technology node is shown in Figs. 8.8 and 8.9, respectively. The remaining process steps are then scaled so that the total number of steps matches those for new technology nodes (Fig. 8.10). Only the major process steps are estimated and included in MEMSCOST, and a number of minor and less expensive steps are not included. The resultant process list is then entered into the *Process Library* for cost calculations.

$$N_{TN} = (Mk_{TN} + \mathrm{Im}_{TN} + m_{TN}) + nm_c \tag{8.3}$$

where:

N_{TN} = number of process steps in selected technology node
Mk_{TN} = number of mask steps in selected technology node

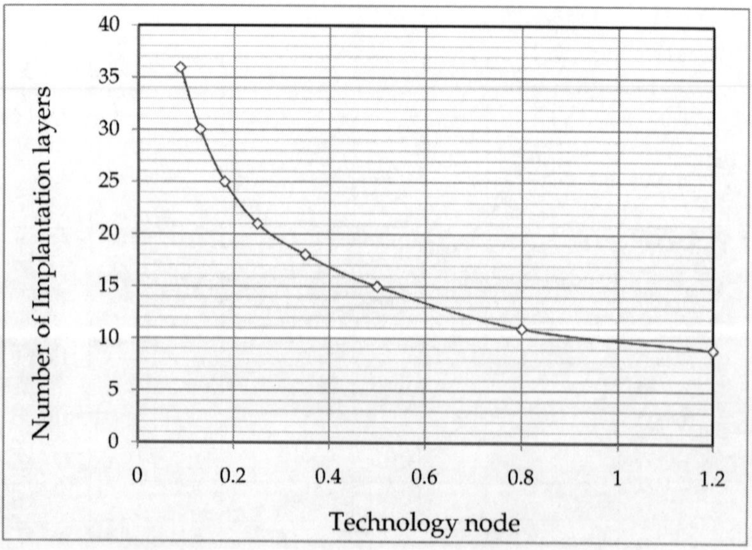

Figure 8.8 Number of implantation layers as a function of the technology node.

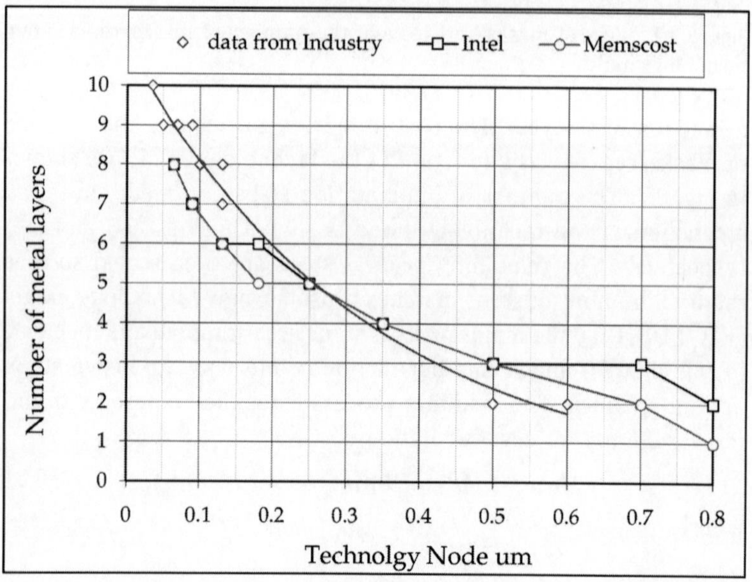

Figure 8.9 Estimated number of metallisation layers per technology node.

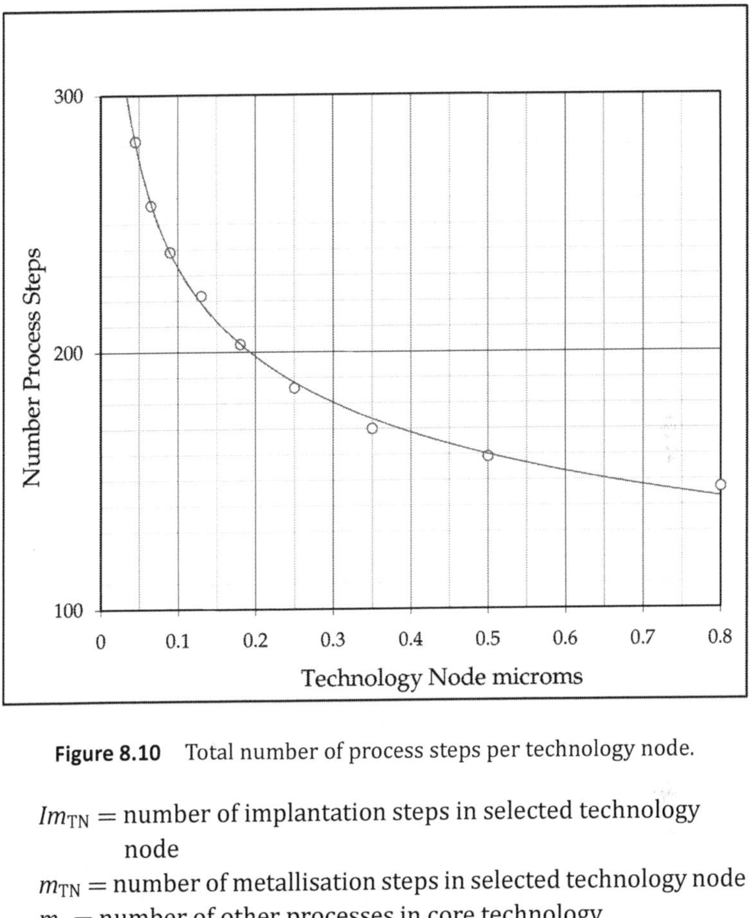

Figure 8.10 Total number of process steps per technology node.

Im_{TN} = number of implantation steps in selected technology node
m_{TN} = number of metallisation steps in selected technology node
m_c = number of other processes in core technology
n = normalising factor to make $N_{0.18} = 200$ steps

The cost estimates resulting from this simplified BiCMOS/CMOS model can be compared with the limited amount of publicly available price data (GSA, 2011), as shown in Fig. 8.11.

The cost of 200 mm diameter processed wafers, tested but unpackaged, is estimated by MEMSCOST for 0.13 μm, 0.18 μm, 0.25 μm and 0.35 μm technology nodes. Assuming these nodes are sold in equal quantities, the average unit cost calculated by MEMSCOST would be \$513. The GSA data gives an average price of \$820, so the estimated profit margin would be 38%.

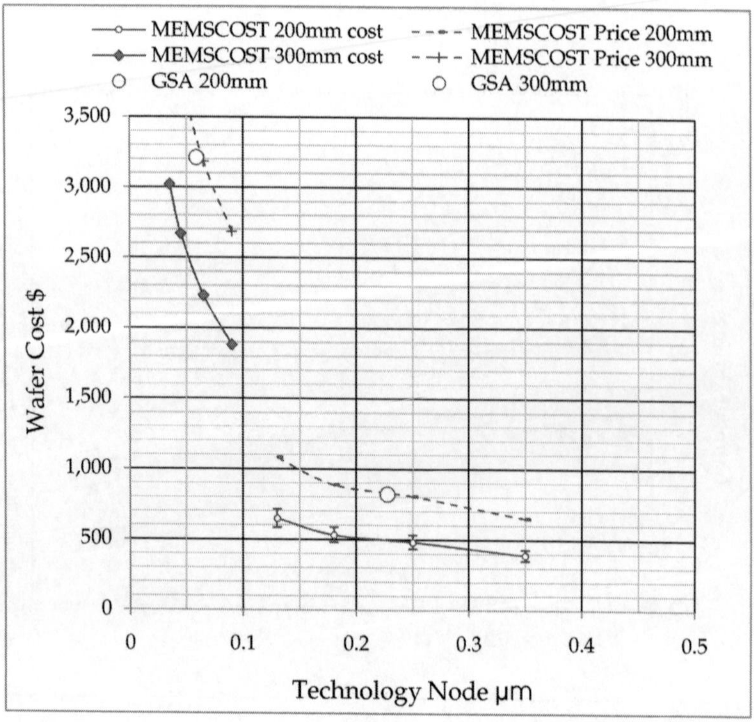

Figure 8.11 Comparison between MEMSCOST estimates for 2011 and the GSA market data for 2011. Annual output is assumed to be 500,000 wafers with an average die size of 10 mm × 10 mm.

The cost of 300 mm diameter processed wafers, tested but unpackaged, is estimated for 0.035 μm, 0.045 μm, 0.065 μm and 0.09 μm technology nodes. Assuming these nodes are sold in equal quantities, then the average unit cost would be $2,448. The GSA data gives an average price of $3,210, so the estimated profit margin would be 24%. These profit margins are calculated for 200 mm (38%) and 300 mm wafers (24%) and are similar to those quoted in the literature (SEMI, 2011; HSBC, 2011). This is sufficiently accurate for MEMSCOST.

Other useful data is publically available. Figure 8.12 shows the reduction in linewidths over the years 1980 to 2010 and the increase in the average die area for memory and for logic. Both the linewidth (L) and the die area (A_D) are strongly related to the year

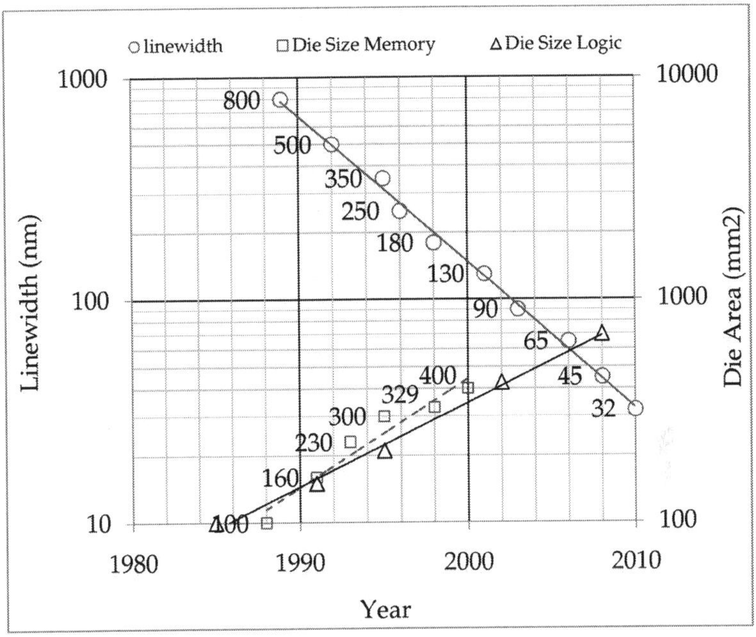

Figure 8.12 CMOS linewidth and die size as a function of the year of introduction.

of introduction to the market (Y_n). Moore's law states, 'The number of transistors in a silicon chip doubles every 18 months'. This can be verified from the data, assuming that the number of transistors is inversely proportional to the linewidth and proportional to the die area. From Fig. 8.12 linewidth is an inverse function of the year ($e^{-0.151Y}$) and the die area, averaged for memory and logic, a function of the year ($e^{-0.1Y}$).

$$(L_{n+1}/L_n)^2 = e^{[-0.15(Y_{e+1}-Y_e)]^2} \tag{8.4}$$

$$(A_{Dn+1}/A_D) = e^{[-0.1(Y_{n+1}-Y_n)]} \tag{8.5}$$

$$(L_n/L_{n+1})^2(A_{Dn+1}/A_{Dn}) = e^{0A(Y_{n+1}-Y_n)} \tag{8.6}$$

Equation 8.6 equals 2 when $Y_{n+1} = Y_n = 1.7$ years, a reasonable confirmation of Moore's law over a 25-year period.

8.8 Financial Data

MEMSCOST financial data is used to calculate the cost per hour of operating each machine used in a process. This data includes the salary and the hours worked per direct staff member. Overheads required to cover non-direct staff may be levelled in a variety of ways. The method chosen in MEMSCOST is to add an overhead rate as a function of direct staff. A simple staff module for a small factory is shown in Table 8.7, which estimates an average salary of $47,083 ($565,000/12) in 2011.

An overhead rate of 150% will add $70,625 to the manpower rate, resulting in a total charge for manpower plus overheads of $117,708. The overhead will be used to fund the administration salaries and other running costs, such as rent, heating and lighting, etc. Table 8.8 shows typical examples of the staff functions that may

Table 8.7 Cost module to estimate average salary for direct staff

	Salary $	Number	Total $
Technician/operator	35,000	8	280,000
Maintenance engineer	45,000	1	45,000
Development engineer	75,000	1	75,000
Quality control	65,000	1	65,000
Operations manager	100,000	1	100,000
Direct staff costs		12	565,000
Average salary/staff member			47,083

Table 8.8 Allocation of 150% overheads to foundry administration

	Number	Salary ($)	Total ($)
Shipping receiving	1	35,000	35,000
Secretary	4	25,000	100,000
Marketing and sales	2	50,000	100,000
Management			
– Operations director	1	150,000	150,000
– Marketing director	1	150,000	150,000
– Chief executive officer	1	250,000	250,000
Admin. manpower	10	55,000	550,000
Other admin. costs			297,500
Management costs			847,500

be charged to overheads (these are the salaries used in MEMSCOST calculations).

Assuming direct staff work for 1,600 hours/year (48 weeks × 5 days/week × 8 hours/day) then an hourly rate as part of the clean-room costs is $73.56/hour/staff member. Typically, a staff member may be able to look after more than one machine, thereby reducing the hourly rate. For example, the hourly rate would be $36.78/hour for a staff member looking after two machines and $24.52/hour for a staff member looking after three machines.

8.9 Wafer and Die Calculations

The *Wafer and Die Calculator* is the front end for the *Device Cost Calculator*, which supplies costs for each piece of equipment. These costs are based on (a) the annual throughput of wafers, (b) the number of times the equipment is used in the production sequence, as specified in the *Process List*, (c) the throughput for the process conditions specified, calculated from the machine simulators and

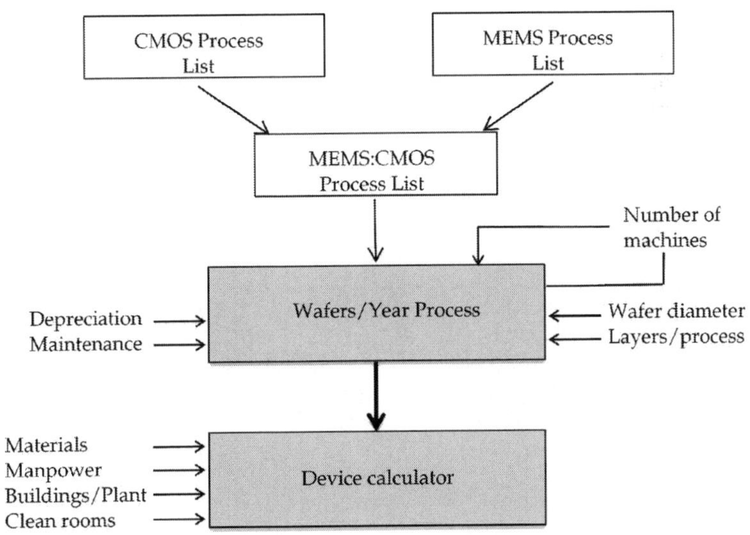

Figure 8.13 Wafer and die inputs to the *Device Cost Calculator*.

(d) the material costs and allowances made for breakdown and preventative maintenance.

An element can be included to cover the cost of previous non-recurrent engineering and attributed on a cost/wafer, cost/hour or percentage-of-the-manufacture basis. MEMSCOST normally adds depreciation costs for the equipment, clean rooms, plant and buildings. Each or all of these charges can be selectively removed depending on different financial policies.

Outputs from the *Device Cost Calculator* are the total cost, the unit cost per wafer and the unit cost per known-good-die (i.e., taking into account the overall process yield). A schematic of the processes is shown in Fig. 8.13 and is performed for fabrication, testing and packaging to get the overall costs. A profit margin can be defined to establish the selling price.

8.10 MEMSCOST Reports

A wide variety of financial and technical information is available from MEMSCOST. Consider a typical MEMS device requiring a cost analysis for both MEMS and CMOS, providing cost data for fabrication, testing and packaging at wafer and die levels.

Table 8.9 summarises the overall technical description, provided by the *Master Set-Up*, of the device being costed. Data includes key aspects of the wafer and die, for example, the wafer and die size,

Table 8.9 Single-die device data

Device data		
Linewidth	0.18	μm
Mask layers	32	
Wafer diameter	200	mm
Die length	1	mm
Die width	1	mm
Wafers/year	10,000	
Dies/wafer (untested)	24,584	
Total dies/year	245,840,000	
Working dies/year (KND)	208,964,000	
Overall yield	85	%

Table 8.10 Cost, profit and selling price summary for a single die MEMS device (as per Table 8.9)

Total cost of wafers and dies	$	%	%
COST/REVENUE			
Fabrication	12,208,42	19.0	32.6
Testing	5,619,045	8.8	15.0
Packaging	19,625,906	30.6	52.4
NRE costs	1,000,000	1.6	0.0
Total	38,452,993	60.0	100.0
Profit margin	25,635,329	40.0	
Selling price	64,088,322	100.0	
PER WAFER		$/cm^2	
Fabrication	1,221	3.88	
Testing	562	1.78	
Packaging	1963	6.23	
NRE costs	100	0.32	
Total	3,845	12.2	
Profit margin	2,564	8.1	
Selling price	6,409	20.3	
PER DIE			
Fabrication	0.0585		
Testing	0.0269		
Packaging	0.0939		
NRE costs	0.0048		
Total	0.1840		
Profit margin	0.1230		
Selling price	0.3070		

Abbreviation: NRE, non-recurring engineering.

wafers/year and yield. This enables the number of good devices per year to be estimated.

The manufacturing costs are reported separately into fabrication, testing and packaging with provision for non-recurrent engineering such as design and prototyping costs. The cost per wafer and per die is calculated, taking into account the yield and profit margin and then the selling price established.

As an example, Table 8.10 shows these costs further sub-divided into total costs, costs per wafer and costs per die. The profit margin is included so that the selling price can be compared with the market surveys.

8.11 Unit Costs

MEMSCOST enables the calculation of unit costs on both a marginal and a fully absorbed basis by allowing the switching in or out of parameters and the direct input of overhead rates. The unit costs calculated and used in this book, however, are on a fully absorbed basis, which captures all costs, including overhead costs and depreciation (see Chapter 12). For the purposes of MEMSCOST, the calculation below assumes that depreciation is in fixed cost and that overheads lie in variable costs.

For any MEMS device, the critical costs will be defined as

$$C_L = C_V N_L + C_F \tag{8.7}$$

The unit cost to process a layer will be:

$$C_L = C_F / N_L + C_V \tag{8.8}$$

where:

C_L = cost to process a layer ($ per layer)
C_F = fixed cost ($ per annum)
C_V = variable cost ($ per wafer layer)
N_L = number of wafer layers per year, that is, number of wafers × number of layers/wafer

The fixed costs per year are calculated from the capital cost of the relevant machines (C_M), the period (n) over which depreciation is calculated and the annual cost of any scheduled maintenance periods and stand-by consumables (M_F), that is,

$$C_F = C_M / n + M_F \tag{8.9}$$

The variable costs/year are calculated from the hourly cost of operating the equipment (C_{OP}), the time to process a layer (T_L) and the cost of throughput-dependent consumables (C_c) per wafer, that is,

$$C_V = C_{OP} T_L + C_C \tag{8.10}$$

T_L is strongly dependent on the throughput of individual machines.

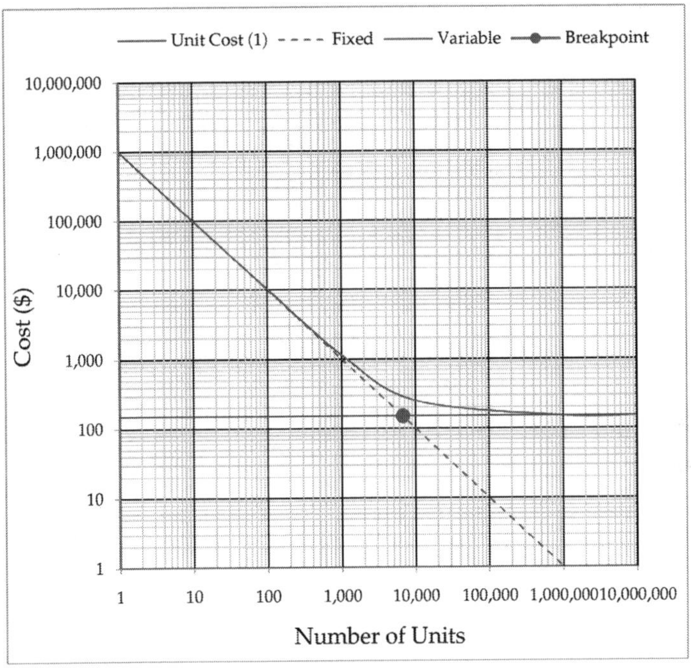

Figure 8.14 Unit costs as function of unit numbers.

For a complete device, *n* processes and equipment costs must be summed; hence

$$\text{Cost} = \sum_{1}^{N} C_V + C_F/N_L \tag{8.11}$$

The form of Eq. 8.11, shown in Fig. 8.14, is fundamental to the cost and price calculations to follow in Chapters 9 and 10 and is calculated as a function of unit numbers (N) for unit costs C_F = \$1,000,000 and C_V = \$150 per unit. For low unit output the unit cost is approximately C_V/N. As output tends to large volumes the costs are asymptotic to C_V. The breakpoint between these two regimes can be defined as $N = C_F/C_V$, that is, 6,667 units in this example.

Unit cost analysis is useful for evaluating the relative cost of individual equipment and processes that make up total manufacturing costs. A simple approach to comparing the cost of the various processing techniques is to consider the cost of using different

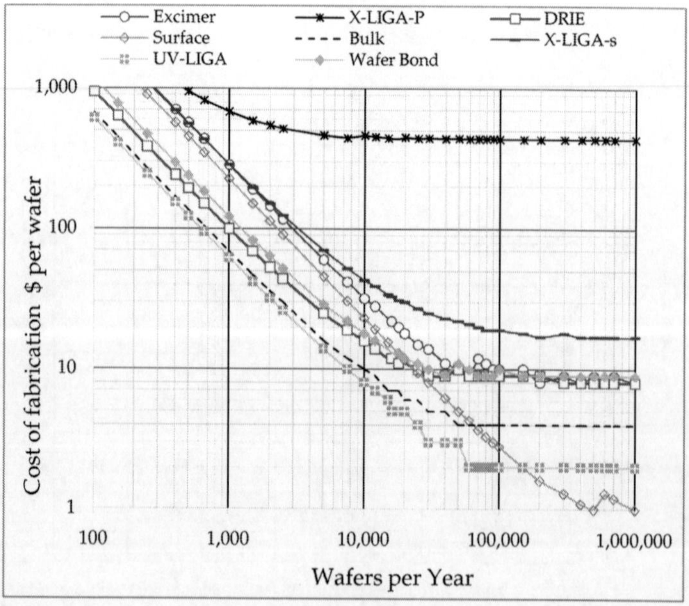

Figure 8.15 Cost of HAR MEMS equipment as a function of wafers/year; 1 mm × 1 mm dies on 200 mm wafers with 100 μm film thickness.

machines to process the same-thickness film on a wafer as the annual output increases. This will take into account the capital cost and the maximum annual output per machine, after which further machines must be purchased.

Figure 8.15 shows the cost of the major HAR processes as a function of wafers per year, assuming a depth requirement into a silicon substrate of 100 μm. The unit costs are calculated according to Eq. 8.8 and hence are a function of both equipment capital cost and throughput. The small perturbations in unit cost occur when additional machines have to be purchased to meet market demand.

X-ray LIGA is the most expensive process, mainly due to the low sensitivity of polymethylmethacrylate (PMMA) to X-rays. However, the use of a much more sensitive resist (e.g., SU-8) brings the unit costs down by an order of magnitude. Eximer laser processing appears to be the next most expensive technique to use due to the relatively slow ablation process. However, it has the advantage of being able to micromachine polymers and other non-silicon materials.

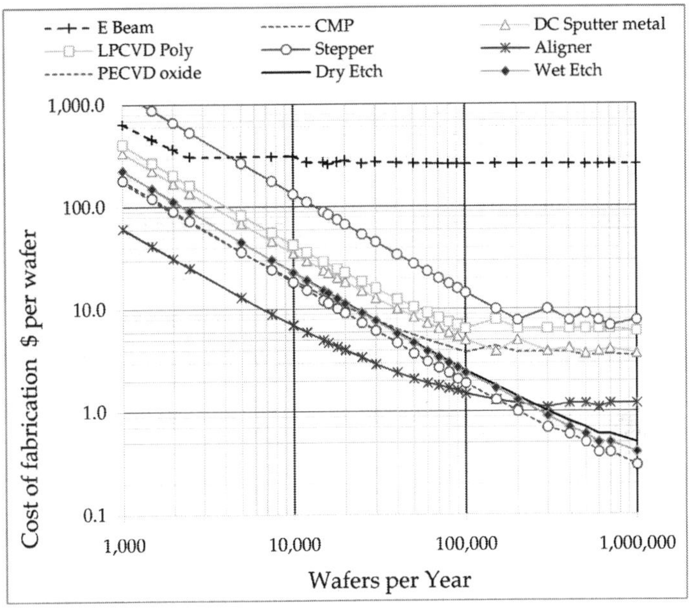

Figure 8.16 Cost of semiconductor equipment as a function of wafers/year; 1 mm × 1 mm dies on 200 mm wafers with 1 μm film thickness.

DRIE and UV-LIGA offer even cheaper HAR processes. DRIE has the advantage of processing silicon structures with near-vertical wall angles and to depths from microns to centimetres. The UV-LIGA process is cheap, requiring only an inexpensive wafer aligner for exposure, and is capable of HAR microstructures with SU-8 and laser radiation, for example, at $\lambda = 248$ nm.

Wafer bonding is a cheap process (the bonding is relatively slow, but the machine is inexpensive), which also offers some interesting unit cost reduction possibilities with multi-wafer architectures.

Bulk micromachining is still used in production and offers the cheapest of the fabrication processes. Electrodeposition is often the method of choice for electroplating, stress-free deep microstructures, which, although a slow process, can be performed with cheap equipment.

Figure 8.16 shows a similar comparison with techniques from semiconductor processing often used to fabricate MEMS devices. A depth of 1 μm is assumed for all the processes shown. Note that HAR processes lie between $10 and $100 per wafer layer for 100 μm layers, while thin-film processes lie between $1 and $10 per wafer layer for 1 μm layers.

The wafer aligner has long been the cheapest lithography tool for MEMS fabrication, particularly for research and relatively low-volume production, but the wafer stepper may become a cost-effective lithography tool as the output requirements approach 1,000,000 wafers per year (as well as meeting likely accuracy and resolution specifications).

The remainder of the processes shown in Fig. 8.16 benefit from batch processing so that a single machine type can meet considerable annual output, for example, 1,000,000 wafers per year. Figure 8.16 also shows why direct-write, Gaussian electron-beam processes are far too expensive to be used in wafer production, either for MEMS or for semiconductors.

References

Allain, M (2010). MEMS cost simulation tool. *Commercialisation of Microsystems Conference, COMS 2010,* Albuquerque, USA.

Grosjean, DE (2007). Reducing defects in integrated surface micromachined accelerometers. micromagazine.fabtech.org/archive/03/03/03/grosjean

GSA (2011). Q1 2011 GSA fabrication pricing report. *Global Semiconductor Alliance.*

HSBC (2011). TSMC financials and valuation. *HSBC Global Research.*

Jones, S (2010). Cost modelling of mems systems. *Commercialisation of Microsystems Conference, COMS 2010,* Albuquerque, USA. http://www.icknowledge.com/our_products/COMS%202010%20rev%202.pdf

Jones, S (2009). Cost modelling as a decision making tool. www.icknowledge.com

IC Knowledge (2012). The IC knowledge—IC cost and price model. www.icknowledge.com/products/icmodel.html

Lawes, RA (2010). Cost analysis—an essential tool for MEMS design and manufacture. *Commercialisation of Microsystems Conference, COMS 2010,* Albuquerque, USA.

Lawes, RA (2009). Cost analysis software for MEMS—a key tool for MEMS commercialization. *Commercialisation of Microsystems Conference, COMS 2009*, Copenhagen.

Lawes, RA (2007), Manufacturing costs for microsystems/MEMS using high aspect ratio techniques. *Microsyst. Technol.*, **13**, 85–95.

SEMI (2011). Global update. Semiconductor Equipment and Materials International.www.semi.org

SystemPlus (2008). Evaluation of MEMS. http://www.systemplus. fr/.../System%20plus%20MEMS%20CEO%202008%20%2presentation.pdf

Chapter 9

Product Costs: Accelerometers

9.1 Introduction

This chapter will assess the accuracy of calculations performed by MEMSCOST using publically available data from microelectro-mechanical systems (MEMS) accelerometer manufacturers and market studies. A representative process list, consisting of the author's estimated process steps and equipment, is used by MEMSCOST to calculate costs and selling prices. This selling price is then reconciled with market data.

An accelerometer measures acceleration when the device is subjected to a change in velocity. The schematic of a single-axis accelerometer in Fig. 9.1 shows the key elements in the MEMS sensor. There is a proof mass whose inertia to acceleration causes the separation of the comb sensor's electrodes (see Chapter 4, Fig. 4.5 for an example) to change along the axis of the acceleration. A spring element limits the movement of the proof mass, by an amount depending on the bending of the flexible beam, and is anchored to the body of the accelerometer. Note that multiple oscillations of the proof mass are damped by viscous friction from gas trapped between the moving mass and the substrate. A voltage

MEMS Cost Analysis: From Laboratory to Industry
Ron Lawes
Copyright © 2014 Pan Stanford Publishing Pte. Ltd.
ISBN 978-981-4411-06-6 (Hardcover), 978-981-4411-07-3 (eBook)
www.panstanford.com

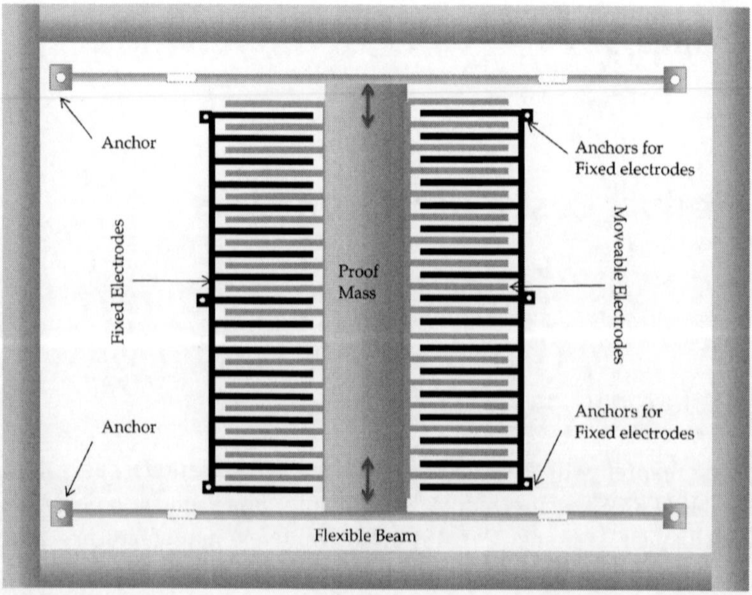

Figure 9.1 Schematic of a single-axis accelerometer (MEMS component only).

proportional to the acceleration is derived by measuring the change in capacitance and hence a change in voltage (see Eq. 9.3).

Considering the electrical properties of the sensor, the total capacitance is

$$C = N_F \varepsilon_0 A/S \tag{9.1}$$

and its sensitivity to motion through small changes δs in a plate separation s is obtained by differentiating Eq. 9.1

$$\delta C = -(N_F \varepsilon_0 A/S^2)\delta s \tag{9.2}$$

where:

C = total capacitance of the sensor (farads)
ε_o = permittivity in vacuum = 8.85×10^{-12} F/m
A = overlapping area of the fixed and moveable electrodes of the sensor (m^2)
s = electrode separation (m)
N_F = number of pairs of fixed-moveable electrodes

The sensitivity of the sensor is proportional to the area between the plates and inversely proportional the square of their separation.

Therefore, for constant charge $Q = CV$

$$\delta V/\delta C = -Q/C^2 = -V_0/C \tag{9.3}$$

where:

V_0 = voltage difference between the fixed and moveable electrodes

Hence by differentiating Eq. 9.3

$$\delta V = (V_0)/C(N_F\varepsilon_0 A/S^2)\delta s \tag{9.4}$$

By substituting Eq. 9.1 for C

$$\delta V = V_0\delta s/s \tag{9.5}$$

The force due to the acceleration of the proof mass must balance the force due to the spring constant of the flexible beam. If the displacement of the moving electrode with respect to the fixed electrode is defined as $x = \delta s$ and damping effects are ignored, then

$$m\delta^2 x/\delta t^2 = kx \tag{9.6}$$

where:

m = proof mass (kg)
k = spring constant (N/m)

Then from Eqs. 9.5 and 9.6

$$\delta^2 x/\delta t^2 = \delta V[ks/(mV_0)] \tag{9.7}$$

that is, the change in voltage across the fixed and moveable electrodes is proportional to the acceleration of the proof mass.

The effect of the flexible beam is to restrain the change in the inter-electrode gap.

The maximum deflection x is at the beam centre, which for a beam clamped at both ends with a central point load F is given by

$$x = FL^3/192EI \tag{9.8}$$

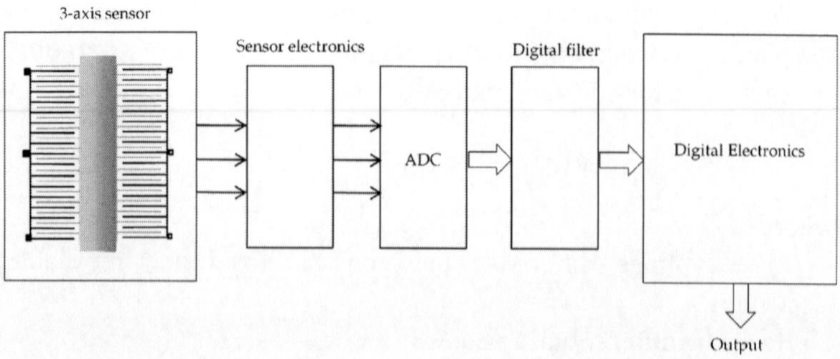

Figure 9.2 Block diagram of the basic signal processing for a digital accelerometer.

where:

E = Young's modulus = 150×10^9 N/m^2
I = moment of inertia = $bd^3/12$ for a beam of breadth b and depth d
$F = m\delta^2 x/\delta t^2$ (Newton's second law of motion)

The modern MEMS accelerometer contains a significant amount of signal processing to achieve high sensitivity and low noise. Figure 9.2 is a block diagram of the basic circuit elements needed to detect the relatively weak signals with low noise. The analogue signals from the sensor are converted into digital form by an analogue-to-digital converter and digitally filtered to reduce noise. Digital signal processing follows to produce the extensive functionality of the leading accelerometers (Analog Devices, 2011).

The signal-processing electronics is in the form of a complementary-metal-oxide semiconductor (CMOS)/bipolar transitor CMOS (BiCMOS)/application-specific integrated circuit (ASIC), which is close to or integrated into the MEMS sensor. The relationship between the sensor and the signal-processing electronics is a distinguishing feature of the construction of a packaged MEMS accelerometer.

Figure 9.3 shows the 'single' construction where the sensor is fabricated in the centre of ASIC die, requiring compatibility between the ASIC and MEMS processing. This approach has been pioneered by Analog Devices through its 'iMEMS' fabrication process (Core

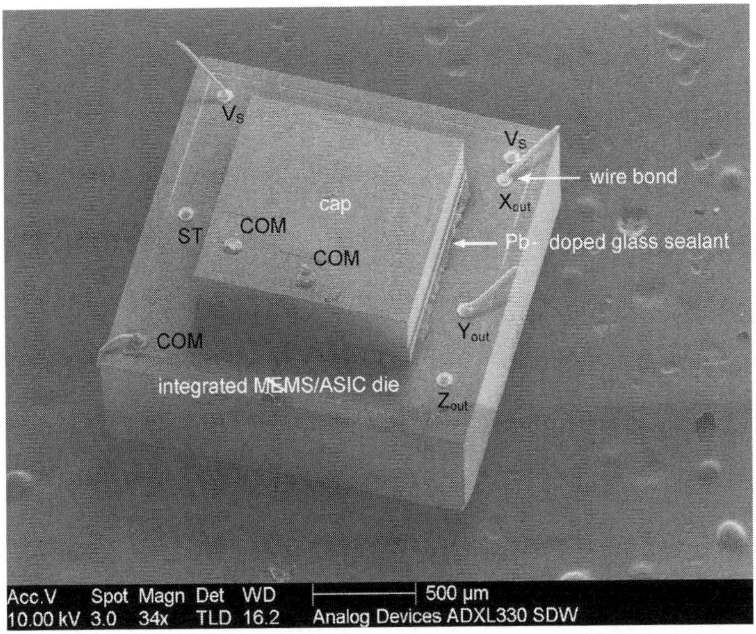

Figure 9.3 Analog Devices ADXL330 accelerometer as single-die design (courtesy Chipworks).

et al., 1993; Bart et al., 1995). Note that the delicate, mechanical MEMS sensor is protected from the ASIC packaging process by fitting a cap over the MEMS chip.

An alternative, and more widely used, construction fabricates the ASIC electronics and the MEMS sensor on separate chips and wire-bonds them together during packaging. The MEMS chip must still be isolated from the rest of the processing by a protective cap. Figure 9.4 shows the construction of ADXL345, manufactured by Analog Devices, where the 'multi-die' form of construction has been adopted, in this case with the ASIC and MEMS dies in a side-by-side configuration. Note that the two dies have been separated from the package for analysis purposes.

Figure 9.5 shows a cross section of the same accelerometer. The dimensions shown are used by MEMSCOST to calculate the wire-bonding costs for a multi-die assembly.

Figure 9.4 Analog Devices ADXL345 accelerometer in a multi-chip design (courtesy Chipworks).

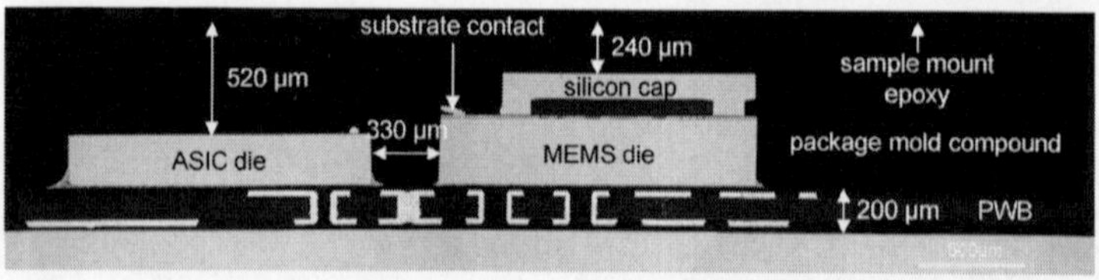

Figure 9.5 Package cross section for the Analog Devices ADXL345 accelerometer (courtesy Chipworks).

Figures 9.6 and 9.7 illustrate variants of the multi-chip construction adopted by the Bosch BMA220, where the MEMS die is mounted on top of the ASIC die in a 'stacked' configuration. The chosen package needs to be able to accommodate the additional height this configuration requires.

The MEMS industry has a wide range of foundries offering fabrication services (see Chapter 11), with the options illustrated in Fig. 9.8. For many years, Analog Devices has used its in-house iMEMS process to manufacture single-die accelerometers (see Fig. 9.3).

Figure 9.6 Bosch BMA220 accelerometer in a multi-chip design (courtesy Chipworks).

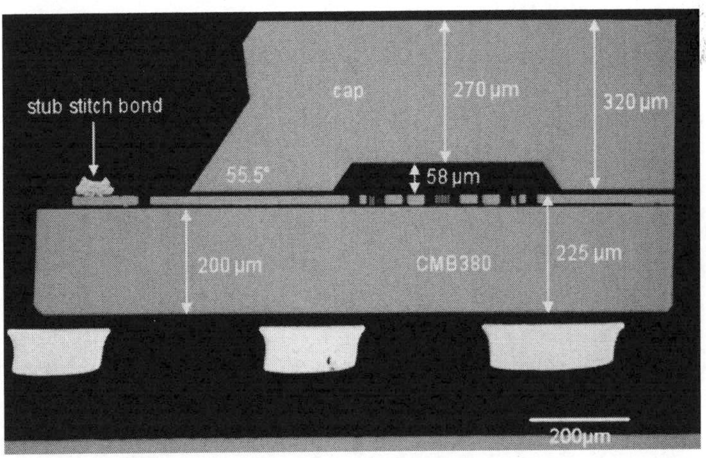

Figure 9.7 Package cross section for the Bosch SMB380 accelerometer (courtesy Chipworks).

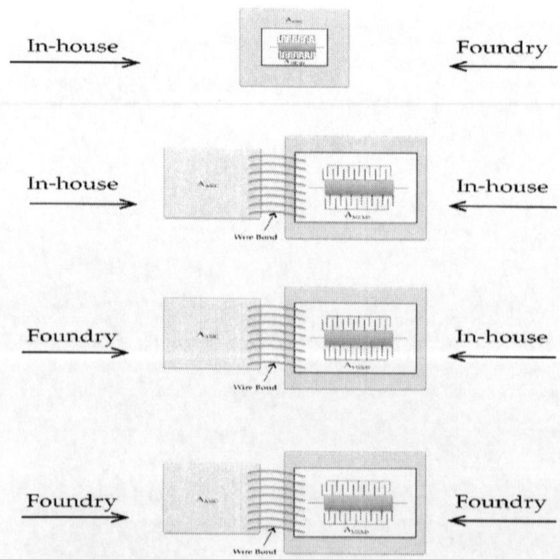

Figure 9.8 Fabrication strategies for single- and multi-die accelerometers.

Other manufacturers also adopted in-house production but with a multi-die assembly.

The logistics of fabrication changed even further when the large-scale integrated circuit (IC) foundries entered the field. For those designs using multi-die configurations, an interesting option arose, where the advantages of purchasing the ASIC die from such a foundry were available, yet the skills of MEMS manufacture, assembly and testing were retained in-house. This is the fabrication methodology adopted by Analog Devices, in 2009, for its ADXL345/346 accelerometers (Dixon & Warren, 2009).

There is also the 'fabless' approach, where the design and possibly the assembly and test are done in-house and the ASIC and MEMS dies are manufactured by a foundry or in separate IC and MEMS foundries. Fabless manufacture obviates the need to meet escalating process development equipment investment. Cost estimates that follow will concentrate on the in-house model, assuming that the cost estimates from the foundry model are less than or equal to those computed by MEMSCOST.

Table 9.1 Wafer and die details for a typical accelerometer-manufacturing process

CMOS BiCMOS	0.18	μm	Dies per wafer	24,584
Mask levels	28		Wafers per year	5,000
Wafer diameter	200	mm	Untested dies	122.9×10^6
Die area	1 mm^2	mm^2	Yield	85%
Scribe lanes	100	μm	Dies/year	104.5×10^6
Exclusion rim	2	mm		
Test regime	Wafer & post-packaging MEMS			
Packaging	Plastic QFN-12	12 pins		

9.2 Costs for Die Fabrication

Wafer and die parameters, for a typical accelerometer, are shown in Table 9.1. The basic fabrication steps for a comb-capacitive sensor in Fig. 9.9 are described below. Fabrication details for the CMOS sensor electronics are not shown.

Step 1 is to deposit thermal oxide and silicon nitride layers to electrically isolate the actuator from the silicon substrate, followed by the deposition of plasma-enhanced chemical vapour deposition (PECVD) silicon oxide as the sacrificial layer. Step 2 patterns and etches the PECVD layer to form the sacrificial layer for the moveable electrode. Low-pressure chemical vapour deposition (LPCVD) polysilicon is then deposited to the thickness needed for the fixed electrode (step 3).

The polysilicon is patterned to define the gaps between the fixed and moving electrodes and etched down to the sacrificial oxide by deep reactive-ion etching (DRIE) (step 4). The use of DRIE is critical to the fabrication of the electrodes, as the electrodes are high-aspect-ratio microstructures. The electrical connections to the electrodes are formed from subsequent sputter deposition and patterning of an aluminium layer. The sacrificial oxide is then etched away (step 5) and critical-point-dried to release the working component. The released comb sensor is shown in its operational state. These main steps, along with a number of detailed steps not shown, are entered into the MEMSCOST *Process Library* for cost analysis.

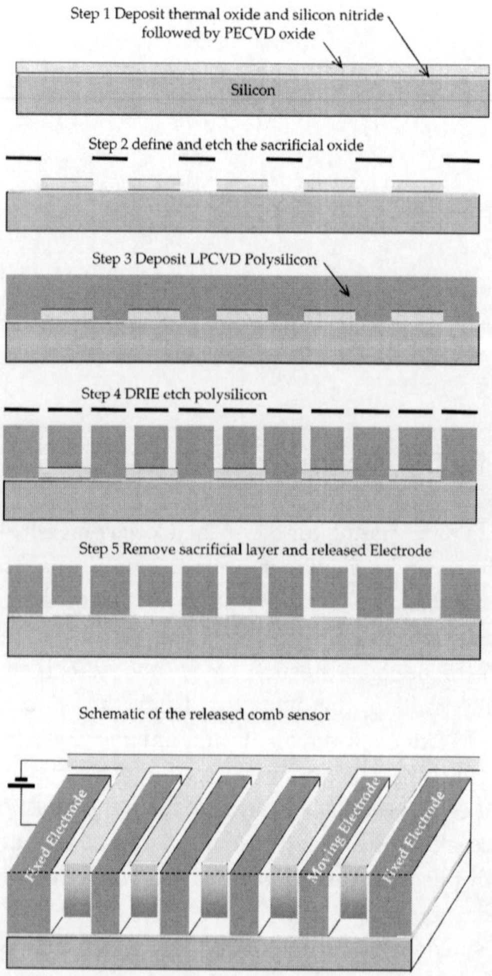

Figure 9.9 Basic fabrication steps for a comb-capacitive sensor.

As expected, the cost of a wafer depends on the size of wafer, the number of wafers per year and the size of the die. As an example, consider the cost of a wafer for a three-axis accelerometer at different wafer sizes and taking into account the different equipment costs, wafer throughput, maintenance and material costs. Figure 9.10 shows the expected increase in wafer costs from 150 mm through 200 mm to 300 mm wafers, assuming an 85% yield ($1,860

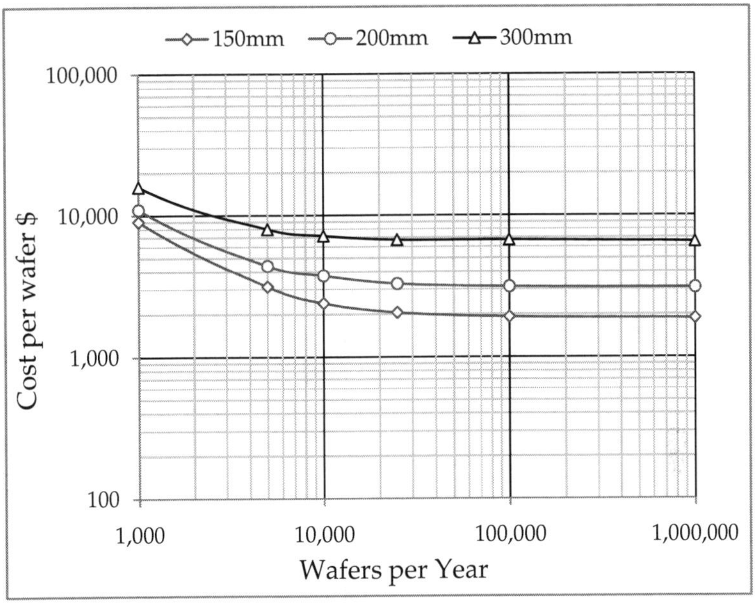

Figure 9.10 Cost of a MEMS wafer (1 × 1 mm die) as a function of wafers/year.

for a 150 mm line, $3,000 for a 200 mm line and $6,500 for a 300 mm line). Figure 9.11 shows the cost of manufacturing a die as a function of wafer output per year, and Fig. 9.12 shows the cost of manufacturing a wafer as a function of die output per year. From Fig. 9.11 it can be seen that there is a reduction in the cost per die arising from a larger wafer size and an increase in the dies per wafer. A 150 mm line could produce tested and packaged dies for a minimum cost of $0.267, the 200 mm line for $0.147, a 45% cost reduction and a 300 mm line for $0.136, a further 7.5% reduction.

The annual number of wafers to be manufactured will depend on the market demand. Figure 9.13 shows the output at which a larger wafer size is more cost-effective than a smaller one so that a 150 mm line can fabricate up to 20 million dies, at a competitive cost, before a 200 mm line becomes an economic possibility. Similarly, once demand goes beyond 200 million dies, a 300 mm line could be considered. However, the cost advantages of larger wafers cannot be realised unless the annual sales are increased.

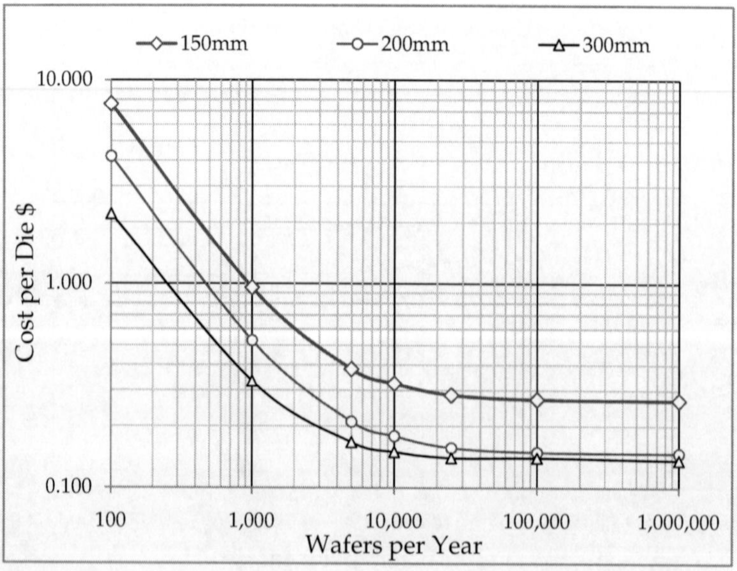

Figure 9.11 Cost of a MEMS die (1 × 1 mm die) as a function of wafers/year.

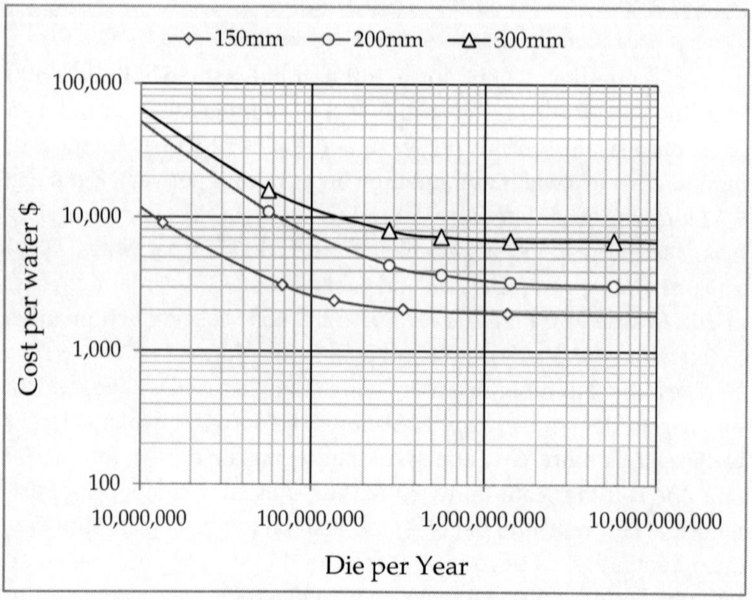

Figure 9.12 Cost of a MEMS wafer (1 × 1 mm die) as a function of dies/year.

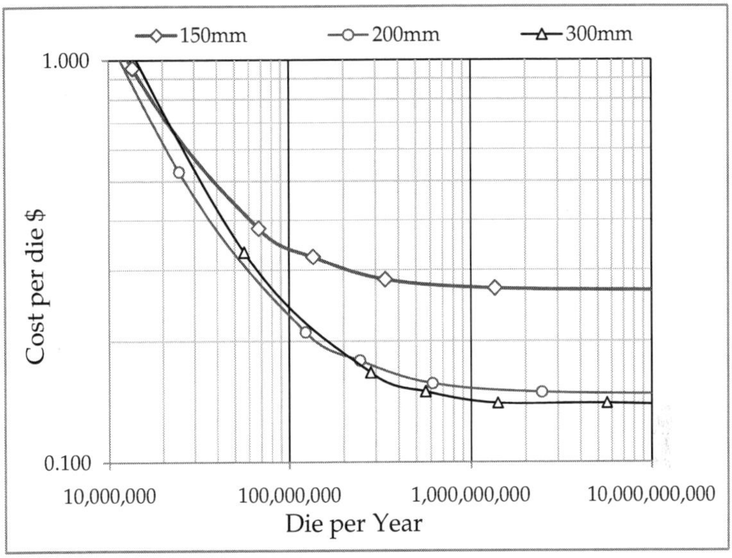

Figure 9.13 Cost of a MEMS die (1 × 1 mm die) as a function of dies/year.

9.3 Revenue Estimates

A good test of the accuracy of the MEMSCOST cost analysis is the comparison with reliable data from commercial sources, and one such source of data is the market survey. Ideally, this would offer a traceable and consistent analysis of the number of units sold per year, the total annual revenue and the unit cost of a product. However, the process by which market survey data is gathered is not widely available, as much of this data is commercially sensitive. The unit costs are usually referred to only in general terms or as a single comment and generally unconnected to a revenue or total output analysis. Nevertheless, careful analysis of the available market data (with a degree of selection to remove anomalies) can reveal useful information.

MEMSCOST estimates the cost of manufacture of an ASIC/MEMS device by determining the reduction of die area from year to year, the number of units sold per year and the annual revenue. The proportion of wafers manufactured on 150 mm and 200 mm wafers is taken into account as production moves progressively from

Table 9.2 Die area for three-axis accelerometers from 2006–2017

Year	Item	Manufacturer	Assembly	ASIC mm^2	MEMS mm^2	Total mm^2
2006	ADXL330	Analog Devices	Single	2.97	0.64	3.61
2006	SMB363	Bosch	Side	3.01	6.64	9.65
2007	LIS331DL	ST	Stack	3.31	3	6.31
2007	SMB380	Bosch	Side	1.7	2.55	4.25
2008	ADXL345	Analog Devices	Stack	2.97	1.84	4.81
2009	ADXL346	Analog Devices	Stack	1.67	1.84	3.51
2009	KXSD9	Kionix	Side	1.94	3.11	5.05
2009	PMMA7660	Freescale	Side	1.92	2.32	4.24
2010	BMA220	Bosch	Stack	1.33	1.37	2.7
2010	LIS2DH/2DM	ST	Stack	1.73	2.03	3.76
2011	ITRS Estimate					2
2013	ITRS Estimate					1
2015	ITRS Estimate					0.5
2017	ITRS Estimate					0.3

Abbreviation: ITRS, International Technology Roadmap for Semiconductors.

150 mm to 200 mm over the period 2006–2017. The test regime is assumed to be post-packaging only, as discussed in Section 9.4.

Table 9.2 lists the ASIC and MEMS die areas of the major three-axis accelerometers introduced in the period from 2006–2011. All except ADXL330 are constructed in multi-die form. This data provides the first input needed for a MEMSCOST estimate. An example of the complexity of the ASIC signal processor is shown in Fig. 9.14 and the associated MEMS die in Fig. 9.15. The ASIC provides advanced power saving, self-test features and a 16-bit digital serial interface (STMicroelectronics, 2012).

Figure 9.16 shows how the combined ASIC and MEMS areas have reduced over the period (Dixon & Warren, 2011; Fraux, 2011; Mounier, 2011). This data includes some predictions for 2011–2017 (ITRS, 2012). An exponential fit to Fig. 9.16 suggests there is a relationship for die area (D_n) as a function of the year of introduction (Y_n), similar to 'Moore's law' for MEMS accelerometers. Solving Eq. 9.9 for an area change by a factor of 2 yields $Y_{n+1} - Y_n = 2.25$ years, that is, 'the area of an accelerometer die halves every 27 months'.

$$D_{n+1}/D_n = e^{-0.308(Y_{n+1}-Y_n)} \tag{9.9}$$

MEMSCOST can generate a simple unit cost from the die area (Fig. 9.16) and from a market survey for total units sold per year. Data on

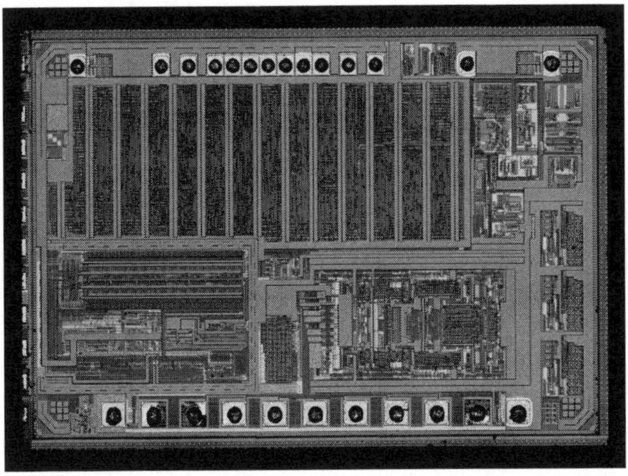

Figure 9.14 ASIC die for the LIS331DLH accelerometer manufactured by STMicroelectronics. The dimensions are 2.18 mm × 1.52 mm. (Courtesy Chipworks.)

Figure 9.15 The LIS331DLH accelerometer MEMS die manufactured by STMicroelectronics. The dimensions are 2.09 mm × 1.43 mm. (Courtesy Chipworks.)

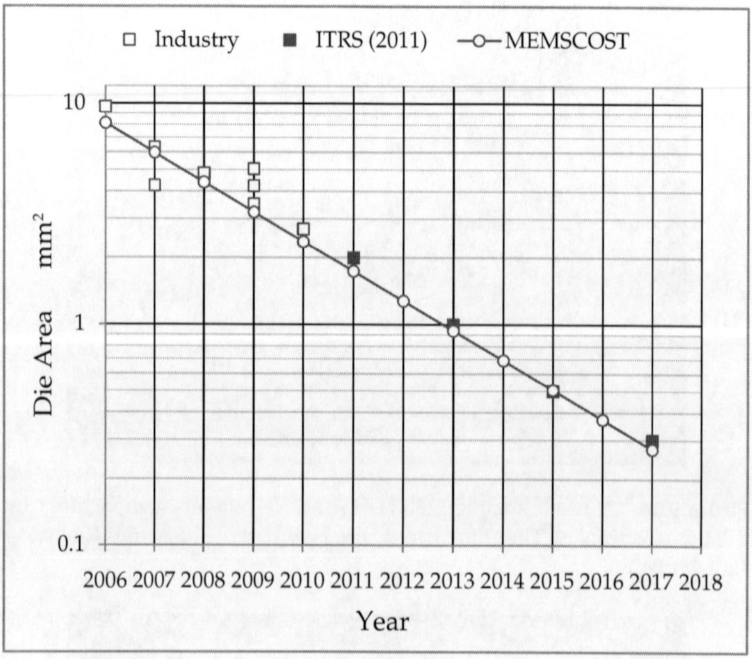

Figure 9.16 Die area for accelerometers from 2006–2017.

the annual number of accelerometers actually sold in past years and estimates for 2012 to 2017 is available from market surveys and is shown in Fig. 9.17 (Basile, 2011).

However, some devices are manufactured on 150 mm lines and others on 200 mm lines. It is assumed that 150 mm foundries will be forced to update to 200 mm wafers and that 200 mm foundries will dominate MEMS production over the next few years. No doubt 300 mm foundries will take over from 200 mm foundries in the future due to cost pressure. The evolution from 150 mm to 200 mm wafers is incorporated into MEMSCOST (Fig. 9.18), with estimates by the author for their respective market shares from 2006–2011 and a prediction for 2012–2017. Note that MEMSCOST assumes that on average, four to six companies divide the market between them. In practice, different manufacturers dominate the market and have an above-average market share in any given year. This gives 'bottom up' estimates for a typical manufacturer's costs from

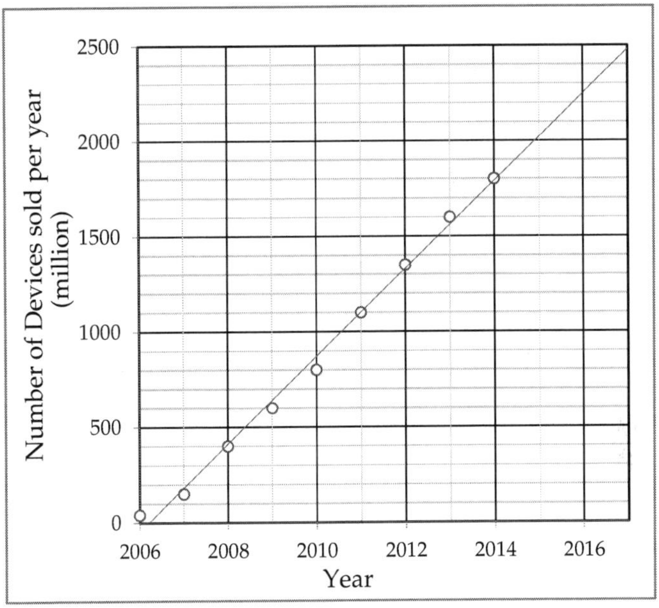

Figure 9.17 Number of accelerometers from 2006–2017 (after iSuppli).

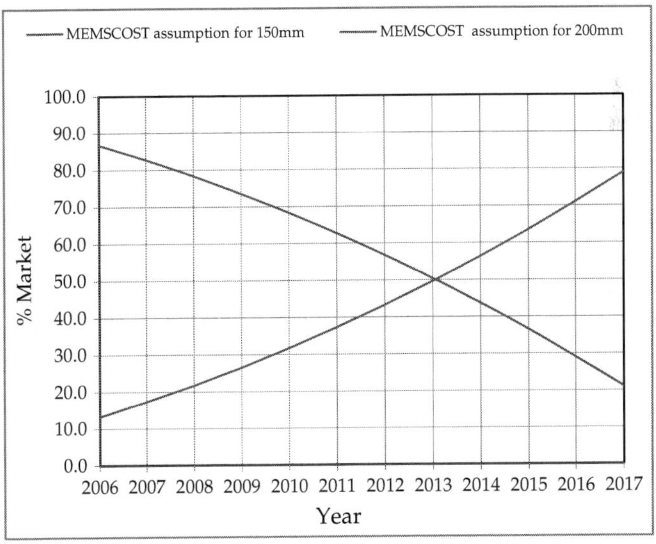

Figure 9.18 Estimated market share of the number of accelerometers by wafer size.

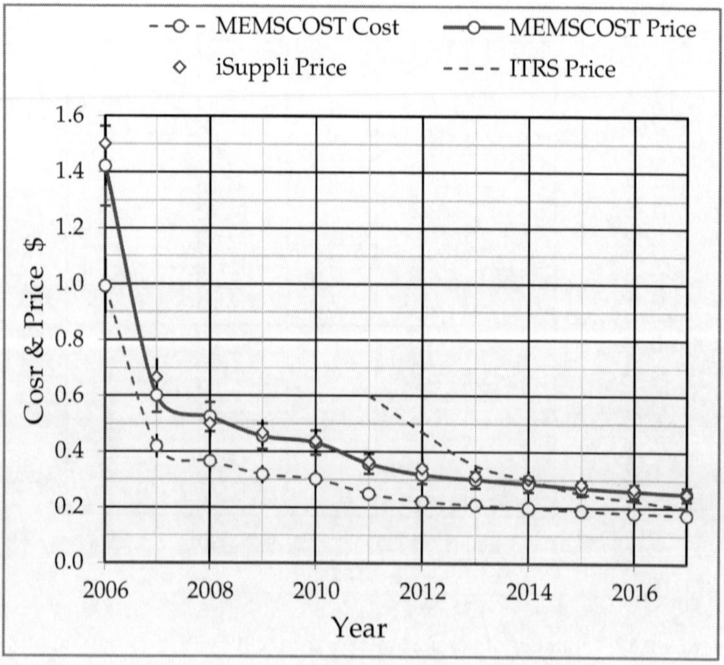

Figure 9.19 Unit costs of MEMS accelerometers (2006–2017).

2006–2017 (see the MEMSCOST cost line in Fig. 9.19). For example, the estimated unit cost to manufacture an accelerometer in 2011 was $0.249.

The revenue derived from the same market survey as before (Basile, 2011) is shown in Fig. 9.20. Dividing the annual revenue by the annual number of units sold (Fig. 9.17) gives an estimate of the market price of an accelerometer. These prices are added to Fig. 9.19 as a 'top down' estimate of price. For 2011 the average unit price is $0.36. The best fit between the MEMSCOST estimated cost and the market survey's price suggests a profit margin of 30%. The MEMSCOST price estimate can now be added to Fig. 9.20, which shows a reasonable fit with the market survey, particularly for later years.

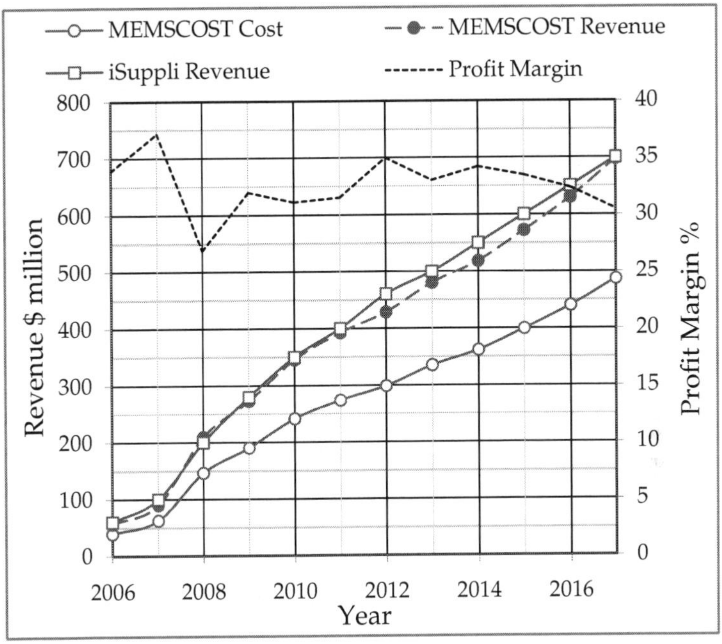

Figure 9.20 Estimates for the revenue from accelerometer sales from 2006 to 2017.

9.4 Cost Comparison for Testing Regimes

There are broadly two approaches to testing, depending on whether the fabrication process is high-yield or low-yield. Where the fabrication process has high yield, most dies tested at the wafer level would be passed to packaging, so it could be cost-effective to omit all pre-packaging wafer tests and test all the packaged devices in a single test. Where the fabrication process is low-yield, it would be costly to package dies without first eliminating the defective dies, so it would be cost-effective to test at the wafer level before packaging, followed by a post-packaging test.

A reasonable assumption is that for a given set of testing and packaging costs, there is a yield above which the first approach is cost-effective and below which the second regime is cost-effective. MEMSCOST can determine the process yields at which either testing regime should be used.

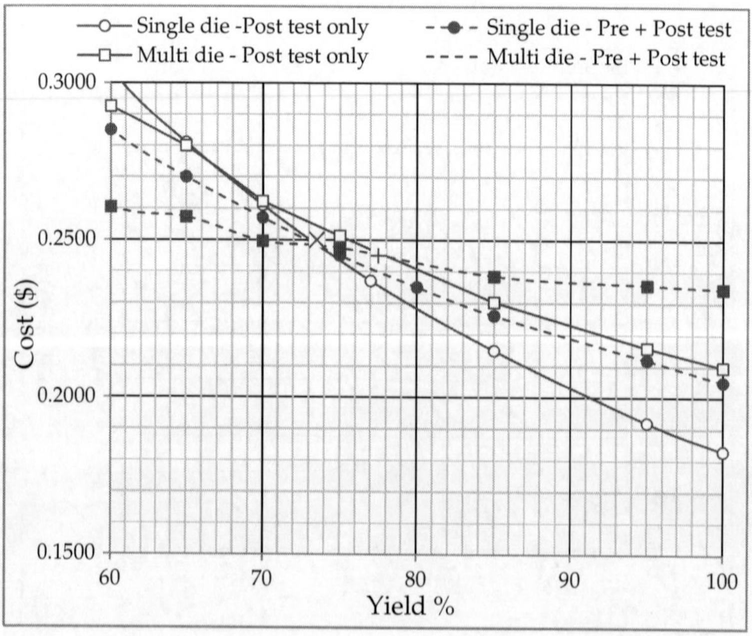

Figure 9.21 Manufacturing costs for pre- and post-packaging tests, compared with post-packaging tests only as a function of yield.

As an example, it is possible to compare the impact of a testing regime on single- and multi-die constructions. For illustration consider two devices, the first a 1 mm × 1 mm accelerometer manufactured as a single-die device on a 200 mm line producing approximately 5,000 wafers per annum and the second the same accelerometer specification but manufactured as a multi-die ASIC/MEMS assembly. In both cases the cost of the accelerometer is calculated as a function of varying process yield.

Figure 9.21 illustrates the two testing regimes, one where the ASIC and MEMS dies are checked at the wafer level and the packaged dies given a final MEMS test and the other where all pre-packaging tests are omitted and only the final MEMS test is performed. For the conditions and processes specified above, there is a process yield at which it is cost-effective to perform only the final MEMS test after packaging. This process yield is 73% for a single-die construction and 77% for a multi-die construction.

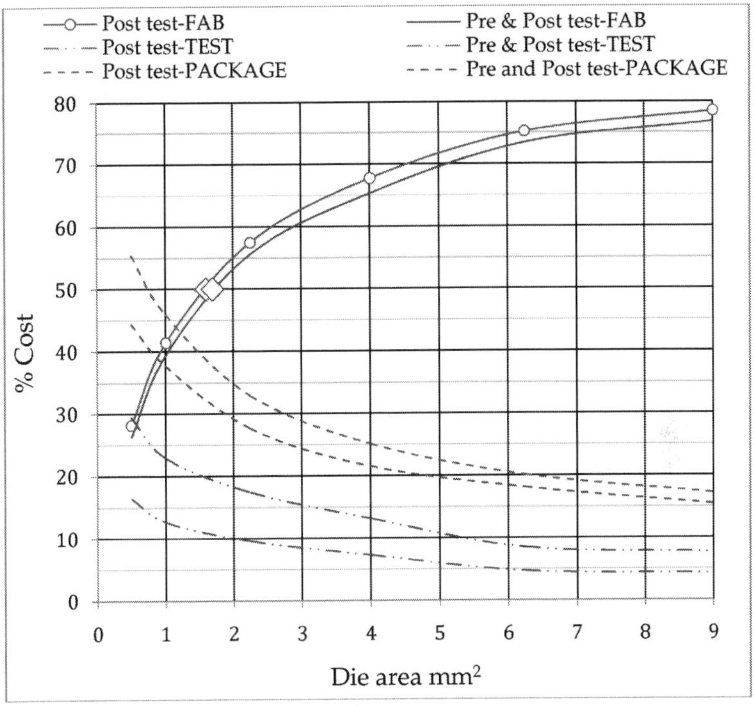

Figure 9.22 Relative costs of fabrication, testing and packaging for changes in die area (see text for assumptions) for a 200 mm line.

The relative cost of device fabrication, testing and packaging for MEMS devices has been the subject of some debate over the last few years, with speculation that packaging can be up to 80% of the cost. Cost analysis shows that the relative costs do not bear a simple relationship to one another but depend on the MEMS type, the die size, the output of wafers and die yield.

Figure 9.22 shows the relationship between fabrication, testing and manufacturing for accelerometers with varying die size, fabricated with an 85% yield on a 200 mm wafer line. If the annual die output has to remain constant, for example, to meet a fixed market, then the relative values between fabrication, testing and packaging will change. As an example, consider a factory with 200 mm line, a fixed wafer capacity (e.g., 5,000 wafers per year) with a fixed yield (85%) but with products with differing die areas. At a die area of 1.6–1.7 mm^2 (e.g., 1.28 mm × 1.28 mm), wafer fabrication represents 50% of the total cost of manufacture, as shown in Fig. 9.22. Therefore testing and packaging combined will only exceed 50% when the die area is reduced below 1.6 mm^2.

9.5 Cost Comparison between Single- and Multi-Die Assemblies

Both the single- and the multi-die design are used to fabricate accelerometers. Assuming that there are no fundamental technical limitations to differentiate between a single- or a multi-die assembly (such as signal levels between the dies, performance, available manufacturing technology or patent protection), the choice of assembly is down to the specification requirements and, potentially, cost. A key question is whether there are significant cost advantages or disadvantages to either approach, which could be the deciding factor in defining the manufacturing and packaging processes.

With a single-die design, the MEMS component is fabricated directly onto the signal-processing ASIC die, where the required space has been left, for example, in the centre of the die (see Chapter 6, Fig. 6.15). In this case, the MEMS fabrication must be compatible with CMOS processing. The advantages for design and construction are that the signal-processing and MEMS components are in close proximity and can be manufactured as an overall process and that only minimal wire bonding is necessary.

With the multi-die product, the signal-processing ASIC is designed and fabricated separately from the MEMS component and may not even be manufactured in the same foundry. The technical advantages of this approach are that the two technologies can be developed and manufactured without any compromise due to the presence of the other. The disadvantages are partly that more silicon is required for two dies and partly that two dies have to be assembled into the package (which may itself have to be larger than for a single die), thus requiring extra wire bonding.

As an example, costs for an accelerometer are analysed assuming a 0.18 μm CMOS/BiCMOS process, producing 100 million, single 1 mm × 1 mm dies on 200 mm wafers, with an 85% yield and

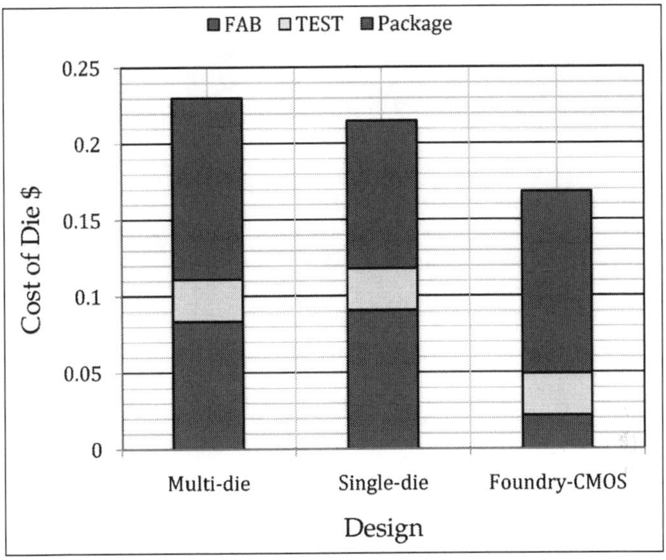

Figure 9.23 Accelerometers—costs for different manufacturing options.

where only post-packaging testing is performed. These costs will be compared with a multi-die assembly, where it is assumed that the ASIC die, designed and fabricated, is 80% of the single-die area and that the separate MEMS dies will be 125% of the multi-die ASIC area, that is, equal to the single-die ASIC area. Both costs are compared with the overall cost of having the ASIC fabricated in an external foundry.

Figure 9.23 shows the unit cost of a fabricated, tested and packaged device for the two methods of in-house manufacture. For the cost structure and values used in this example, the multi-die design costs 7% more, although cost of manufacture may not be the only determining factor (see above).

As overall costs and performance drive the reduction in die size, a company may not be able simply to change from multi- to single-die construction due to the large investment in resources that would be required. In recent years specialist CMOS/BiCMOS foundries have exploited the hybrid construction of accelerometers by supplying the ASIC, while the integrated device manufacturer (IDM) manufactures the MEMS sensor, packages the two dies

Table 9.3 Accelerometers—manufacturing options

	IDM single	IDM multi	ASIC foundry	
ASIC die area	1	0.44	0.44	mm^2
MEMS area	–	0.56	0.56	mm^2
ASIC cost	0.0906	0.0725	0.0066	$
Foundry profit margin	–	–	40	%
ASIC cost	–	0.0725	0.0110	$
Fabricate sensor, test and package at IDM				
Sensor (fabrication)	–	0.0111	0.0221	$
Test	0.0273	0.0273	0.0273	$
Package	0.0971	0.1194	0.1194	$
Cost	0.2150	0.2303	0.1688	$

and undertakes final testing. Cost analysis can reveal the possible financial benefits that can arise from use of an external foundry to manufacture the signal-processing ASIC. A specialist ASIC foundry will have the benefit of very large volumes of CMOS/BiCMOS circuits, for example, 1 million wafers per year, and should be able to offer a much reduced price compared with an in-house manufacturer, which produces only sufficient ASICs for the accelerometers. The ASIC foundry would be able to offer a part of its output at a competitive price with a reasonable profit.

Consider a major 300 mm, 1,000,000 wafers per year, pure-play ASIC foundry, with a profit margin of 40%, providing untested ASIC dies to an accelerometer manufacturer. Table 9.3 shows that the cost to the accelerometer manufacturer would be reduced to $0.169 compared with the cost to a single-die in-house design of $0.215, a reduction of 21.6%. However, the manufacturer of a multi-die design that costs $0.230 could reduce costs by 26.8% through the use of an external foundry for the ASIC signal processor, while maintaining construction of the MEMS component in-house, along with testing and fabrication.

The methodology can be extended to dies of any size and therefore give some guidance as to the most cost-effective option for the manufacture of complex MEMS devices. Figure 9.24 shows that at a die area of 2.6 mm^2 the relative costs of single- and multi-die designs cross over, with the single-die design more cost-effective below and the multi-die design more cost-effective above this area.

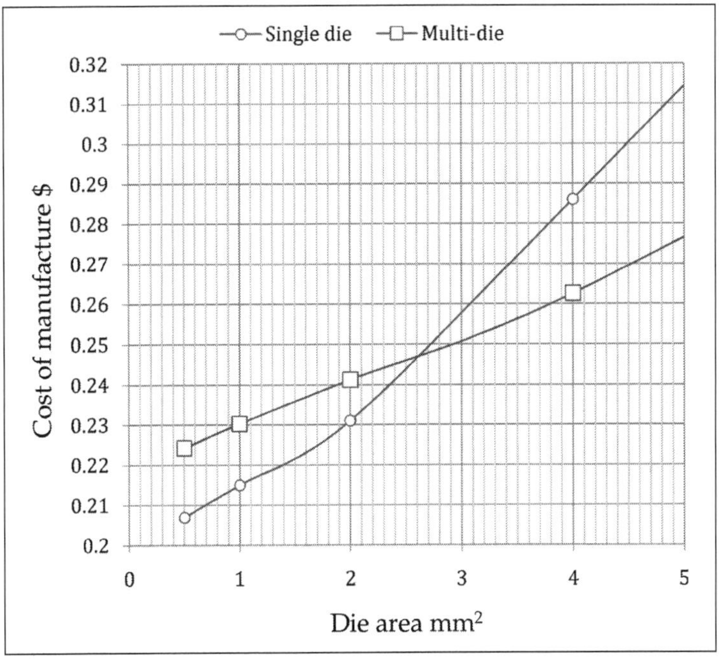

Figure 9.24 Cost comparison between single- and multi-die design as a function of the die area.

9.6 Sensitivity of Costs to Manufacturing Parameters

MEMSCOST can provide an overall analysis of the sensitivity of accelerometer costs to changes in the various financial and manufacturing parameters. By making small changes in a given parameter (x), MEMSCOST can calculate the change in cost (C_A) and hence the differential coefficient (δ_{coeff}). Table 9.4 shows the results for a number of differential cost coefficients. The net affect on cost (Eq. 9.10) can be determined from the differential cost coefficients by the estimated or known change in a given parameter (δx), that is,

$$\delta_{coeff} = \delta x(\delta C_A/\delta x) \tag{9.10}$$

For example, the differential coefficient for process yield is -0.0015 $/% yield at an average process yield of 85%. If this yield is changed by 10% then $\delta x = 8.5\%$ and the effect on overall cost is $0.0119 (8.5 $\times$ 0.0014). Table 9.4 shows similar calculations,

Table 9.4 Differential cost coefficients and % change of overall cost

Parameter	Coefficient	Nominal	Change	10%	$	%
Yield	-0.0015	$/%	85 %	8.5%	0.0119	5.5
Machines/ technician	−0.052	$/machine/ technician	2	0.2	0.0104	4.8
Wafers/year	−8.50E-10	$/wafer/year	1.0E+08	1.0E+07	0.0085	4.0
Staff hours	−5.00E-05	$/hour	1600	160	0.0080	3.7
Depreciation	−0.0115	$/year	5	0.5	0.0058	2.7
Uptime	−0.00015	$/%	90%	9	0.0014	0.6
Capital package	2.00E-05	$/%	100%	10	−0.0002	−0.1
Maintenance	0.0025	$/%	2	0.2	−0.0005	−0.2
Capital test	1.00E-04	$/%	100%	10	−0.0010	−0.5
Die area	0.0160	$/mm^2	1	0.1	−0.0016	−0.7
Capital fab.	3E-04	$/%	100%	10	−0.003	−1.4
Tech. node	−0.0035	%/μm	0.18	1 node	−0.0041	−1.9
Overheads	0.0003	$/%	150%	15	−0.005	−2.3
Salary	8E-04	$/$	100%	10	−0.008	3.7

assuming a +10% change in each parameter, and Fig. 9.25 shows parameters that decrease costs (shaded green) and those that increase costs (shaded red). The main parameters leading to a reduction in costs are yield, the number of machines serviced per technician, the wafers per year, staff hours worked and the depreciation period. The main parameters leading to an increase in costs are technician salaries, overheads and the technology node for the ASIC signal processor.

Assuming all the parameters were in error by +/−10% then Eq. 9.11 can provide an estimate for the uncertainty (δ) in calculating the unit cost.

$$\delta = \sqrt{\sum (\delta C_A/\delta x)^2 \delta x^2} \tag{9.11}$$

Applying the parameters in Table 9.4 to this equation results in an estimated uncertainty of +/−10.8%, that is, δ = \$0.0232. The estimated overall cost of a hybrid accelerometer is \$0.215 manufactured with the parameters shown. If δ is a measure of 1 standard deviation of error, then the unit costs could be assumed to lie between \$0.169 and \$0.261 (2 standard deviations).

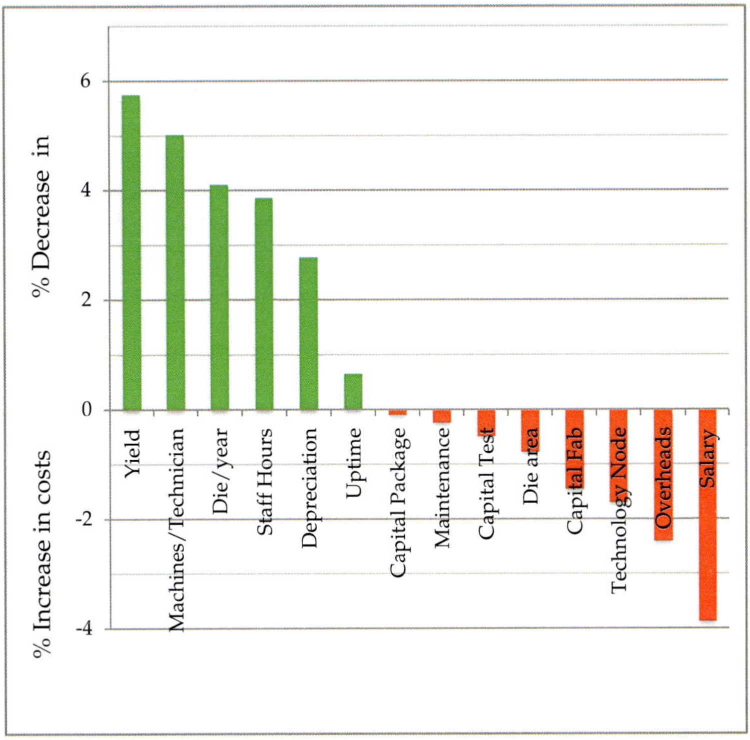

Figure 9.25 Cost sensitivities—% change in overall costs per parameter.

References

ADI (2012). Micromachined products division. www.analog.com/en/content/mpd/fca.html.

Analog Devices (1012). ADXL345: 3 axis digital accelerometer. www.analog.com/en/mems-sensors/adxl345/products/product.html

Bart, S, et al. (1995). Design rules for a reliable surface micromachined IC sensor. *1995 IEEE Reliability Phys. Conf., Las Vegas*, 311–317.

Basile, A (2011). MEMS move into Sart sensors: the STMicroelectronics iNEMO ecosystem. www.mms2011.org/Presentation/09-Basile.pdf

Core, TA, Tsang, WK, and Sherman, SJ (1993). Fabrication technology for an integrated surface micromachined sensor. *Solid State Technol.*, **31**, 39–47.

Dixon-Warren, St, J (2009). Chipworks: inside Analog Devices' new MEMS strategy. Memsblog.wordpress.com/2009/09/28/chipworks.

Fraux, R (2011). MEMS inertial sensors in consumer electronics. *i-micronews*, April(6).

Fraux, R (2011). STMicroelectronics innovation in wafer bonding techniques shrinks MEMS die size and cost. *i-micronews*, November(21).

ITRS (2011). Micro-electro-mechanical systems. *International Roadmap for Semiconductors 2011 Edition Micro-Electro-Mechanical Systems.*

Mounier, E, and Laurent, R (2011). Technology trends for inertial MEMS. *Yole Development Report.*

STMicroelectronics (2012). LIS331DLH MEMS motion sensor: ultra low power high performance 3 axis digital accelerometer. www.st.com/internet/analog/product/218132.jsp

Chapter 10

Product Costs: Microphones

10.1 Introduction

This chapter will assess the accuracy of calculations performed by MEMSCOST using publically available data from microelectromechanical systems (MEMS) microphone manufacturers and market studies. A representative process list, consisting of the author's estimated process steps and equipment, is used by MEMSCOST to calculate costs and selling prices. This selling price is then reconciled with market data.

A MEMS capacitive microphone consists of a variable capacitor with a flexible diaphragm acting as one electrode and one or two rigid backplates forming the other electrode(s). As the diaphragm flexes in response to acoustic pressure, it causes a change in capacitance that can be detected, amplified and converted to a digital signal. Figure 10.1 shows the configuration for a single backplate microphone and Fig 10.2 that for a dual backplate microphone. The diaphragm is situated above an air filled cavity, which acts as an energy store.

The MEMS microphone can be fabricated in silicon-based technology and is therefore compatible with complementary-metal-oxide semiconductor (CMOS) processes. The overall chip size of a

MEMS Cost Analysis: From Laboratory to Industry
Ron Lawes
Copyright © 2014 Pan Stanford Publishing Pte. Ltd.
ISBN 978-981-4411-06-6 (Hardcover), 978-981-4411-07-3 (eBook)
www.panstanford.com

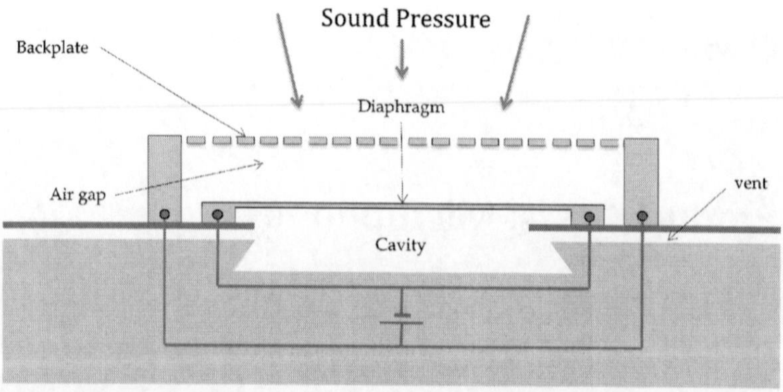

Figure 10.1 A single backplate MEMS microphone.

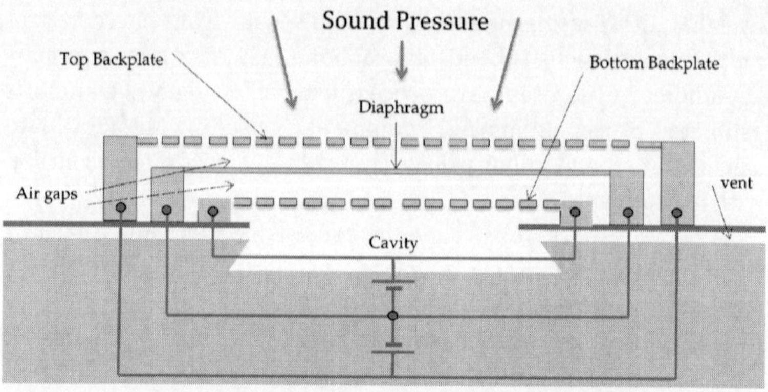

Figure 10.2 A dual backplate MEMS microphone.

typical MEMS microphone is approximately 2 mm × 2 mm with a membrane diameter of 1 mm (see later in Fig. 10.13 for trends in dimensions). The finished product offers excellent thermal and humidity stability. Figure 10.3 illustrates a typical front-end charge amplifier for a single backplate microphone. Note that the direct current (DC) bias circuitry is not shown.

Consider a single backplate microphone with the following parameters

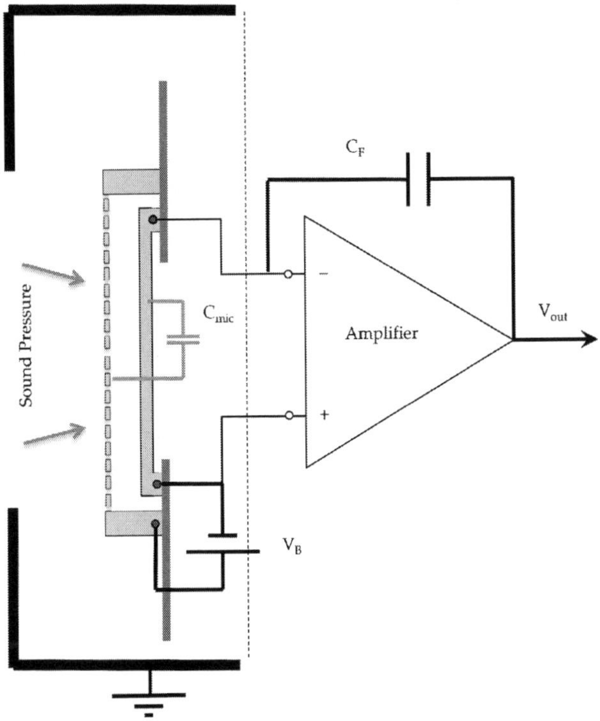

Figure 10.3 A charge amplifier configuration for signal processing a MEMS microphone. C_{mic} is a notional element indicating the capacitance of the microphone.

C_{mic} = microphone capacitance (farads)
Q = charge on microphone capacitor plates (coulombs)
V_B = constant bias voltage (volts)
s = separation of plates (m)
A = area of microphone diaphragm (m^2)
ε_o = permittivity in vacuum (F/m)

and a charge amplifier where

V_{out} = amplifier output voltage
C_F = feedback capacitor (farads)

$$Q = CV \tag{10.1}$$

Therefore, at constant voltage, the change of charge in the microphone, due to acoustic pressure on the diaphragm, is by differentiation of Eq. 10.1.

$$\delta Q = V_{\mathrm{B}}\delta C_{\mathrm{mic}} \quad (10.2)$$

The microphone capacitance C_{mic} is given by

$$C_{\mathrm{mic}} = A\varepsilon_0/s \quad (10.3)$$

and by differentiation of Eq. 10.3

$$\delta C_{\mathrm{mic}} = -A\varepsilon_0\delta s/s^2 = -C_{\mathrm{mic}}\delta s/s \quad (10.4)$$

Combining Eqs. 10.2 and 10.4

$$\delta Q = -V_{\mathrm{B}}(C_{\mathrm{mic}}/C_{\mathrm{F}}\delta s/s) \quad (10.5)$$

Considering the differential amplifier in Fig. 10.3, the charge δQ_{mic}, which is injected into the summing point of the amplifier, is equal to the charge in the feedback capacitor ($V_{\mathrm{out}}C_{\mathrm{F}}$). From Eq. 10.5 above

$$V_{\mathrm{out}} = -V_{\mathrm{B}}(C_{\mathrm{mic}}/C_{\mathrm{F}}\delta s/s) \quad (10.6)$$

As an approximation, δs equals the maximum deflection δ_{max} at the centre of a circular membrane clamped around its periphery and under equal pressure across its surface. This deflection is given by Eq. 10.7 (Schomburg, 2011):

$$\delta_{\mathrm{max}} = 3P_{\mathrm{A}}r4(1-v^2)/(Et^3) \quad (10.7)$$

where

P_{A} = uniform pressure change from acoustic input (pascals)
r = radius of a circular membrane (m)
ν = Poisson's ratio (0.3 approximately for silicon)
E = Young's modulus (150 gigapascals for silicon)
t = thickness of diaphragm/membrane (m)

Combining Eqs. 10.6 and 10.7

$$V_{\mathrm{out}} = 3(V_{\mathrm{B}}/s)(C_{\mathrm{mic}}/C_{\mathrm{F}})P_{\mathrm{A}}r4(1-v^2)/(Et^2) \quad (10.8)$$

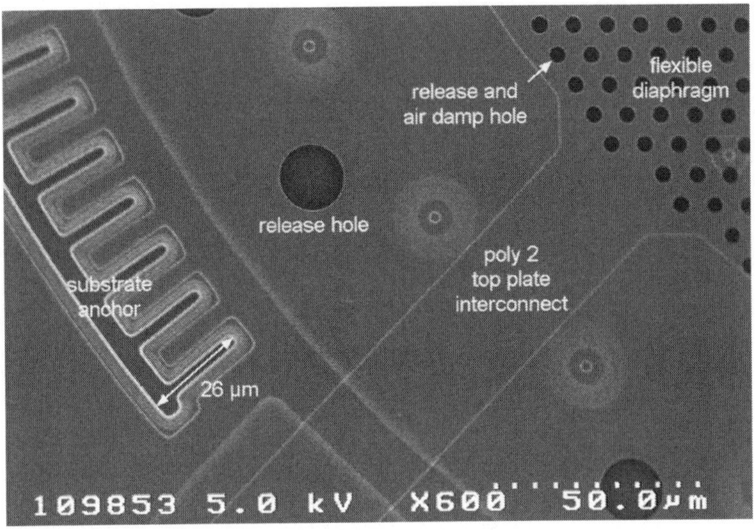

Figure 10.4 Knowles S2.14 MEMS microphone die (courtesy Chipworks).

Hence, V_{out} is proportional to the acoustic pressure and inversely proportional to the inter-electrode separation. It is proportional to the fourth power of the diaphragm radius and inversely proportional to the cube of the diaphragm thickness. Similar working for the double backplate microphone shows that the change in capacitance is double that of the single backplate microphone and hence the amplifier output voltage is doubled.

Some details of a commercial dual backplate MEMS microphone are shown in Figs. 10.4–10.7. Figure 10.4 is a plan view of the Knowles S2.14 microphone, illustrating the construction of the flexible diaphragm and the holes that provide damping. Figure 10.5 is a side view of the microphone, clearly showing a section of the top and bottom backplates and the diaphragm (compare with the schematic in Fig. 10.2). The application-specific integrated circuit (ASIC) and MEMS dies for this microphone are shown in Figs. 10.6 and 10.7, respectively. The ASIC and MEMS dies are separate, that is, a multi-die configuration.

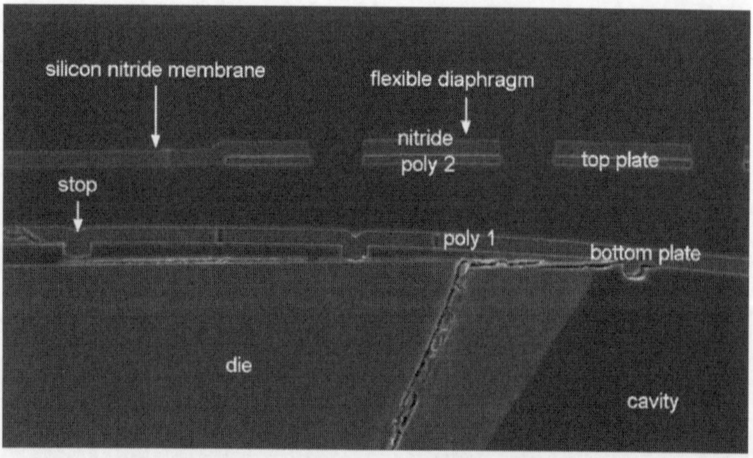

Figure 10.5 Knowles S2.14 MEMS microphone die (courtesy Chipworks).

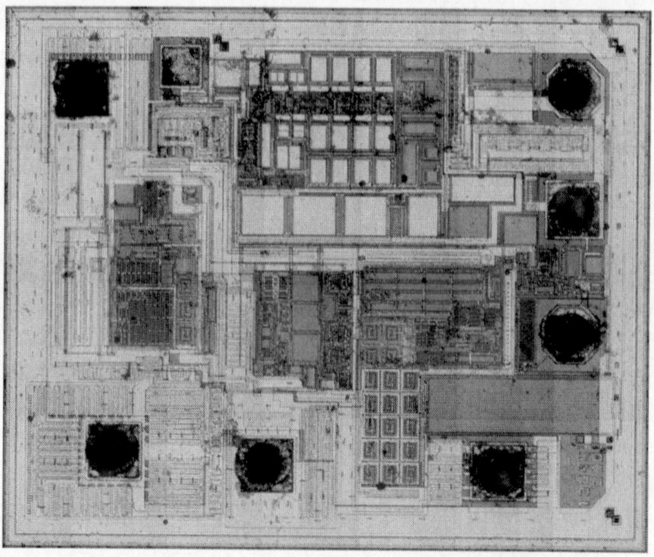

Figure 10.6 Knowles S2.14 ASIC microphone die. Dimensions are 1.22 mm × 1.01 mm. (Courtesy Chipworks.)

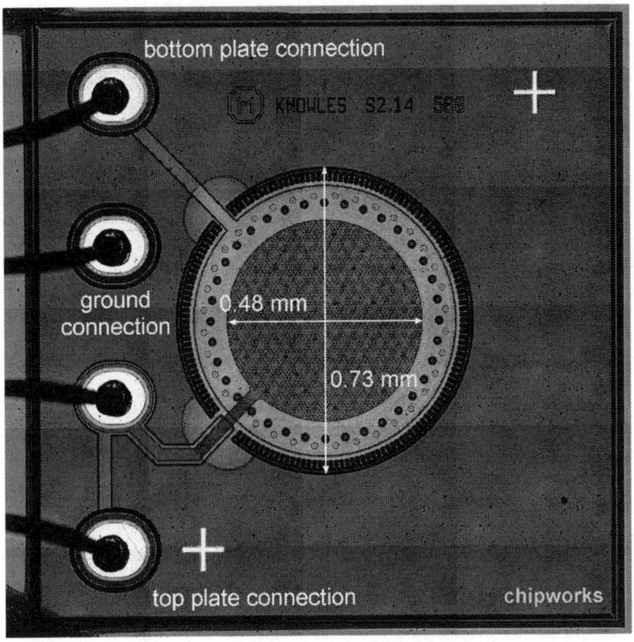

Figure 10.7 Knowles S2.14 MEMS microphone die (courtesy Chipworks).

10.2 Costs for Die Fabrication

The basic fabrication steps to produce the MEMS component of a notional double backplate microphone are shown in Fig. 10.8.

The first steps deposit thermal oxide/silicon nitride layers to electrically isolate the actuator from the silicon substrate (step 1), and then a layer of low-pressure chemical vapour deposition (LPCVD) polysilicon is deposited (step 2). This is patterned to form the electrical connections (the polysilicon is shown in a different colour to emphasise its role).

A sacrificial layer of plasma-enhanced chemical vapour deposition (PECVD) silicon oxide is then deposited (step 3) to distance the subsequent membrane layer from the substrate surface. A layer of LPCVD polysilicon is deposited (step 4) on top of the sacrificial layer, to the thickness required for the bottom backplate, and then patterned to form the necessary perforations.

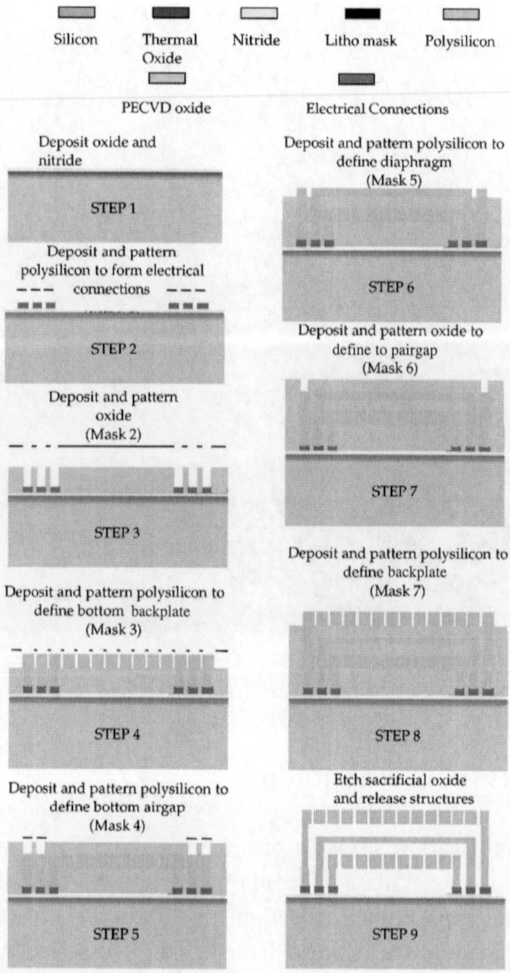

Figure 10.8 Basic process steps for a double backplate MEMS microphone.

A second sacrificial layer of PECVD silicon oxide is deposited (step 5), on top of which is deposited another layer of LPCVD polysilicon (step 6) to a thickness required for the flexible membrane. A third sacrificial layer of PECVD silicon oxide is deposited (step 7), and then an LPCVD polysilicon layer is deposited (step 8) to form the top backplate and patterned to fabricate the perforations. The sacrificial oxides are etched away (step 9) and critical-point-dried

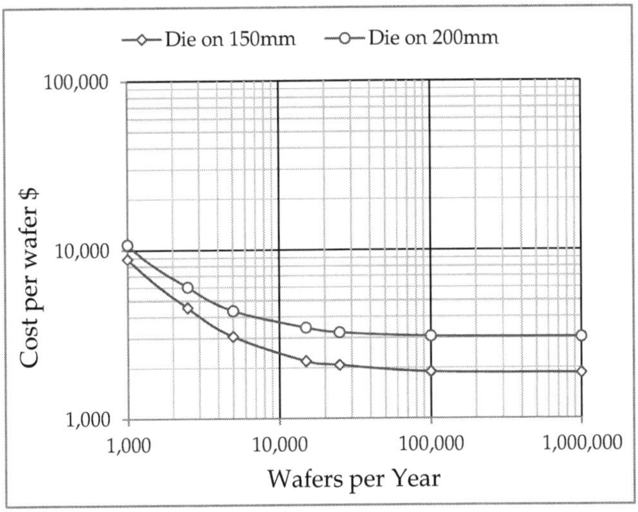

Figure 10.9 Unit cost of a wafer as a function of wafers per year.

to release the flexible polysilicon membrane between the top and bottom backplates.

These fabrication steps, and a number of steps not described here, are entered into the MEMSCOST *Process Library* for cost analysis. Note that Fig. 10.8 only shows the electrode assembly of the microphone. The ASIC signal processor will surround the MEMS microstructure if the microphone is manufactured as a single die. If manufactured as a multi-die microphone then the MEMS die will be much larger to accommodate the wire bond pads connecting the MEMS component to the ASIC die (see earlier Fig. 10.7). In both cases the delicate MEMS structures will be sufficiently removed from the scribe lanes to avoid damage during dicing.

Figures 10.9 to 10.12 show costs as a function of the annual wafer and die output, with the characteristic minimum cost for large volumes. As a function of wafers per year, Fig. 10.9 suggests that the decrease in the cost of wafers slows down beyond 10,000 wafers per year to a minimum cost of $1,860 for a 150 mm line and $3,000 for a 200 mm line. Figure 10.10 shows similar data for the cost of a die at $0.161 and $0.145 for a 150 mm and a 200 mm line, respectively, assuming a 1 mm × 1 mm die.

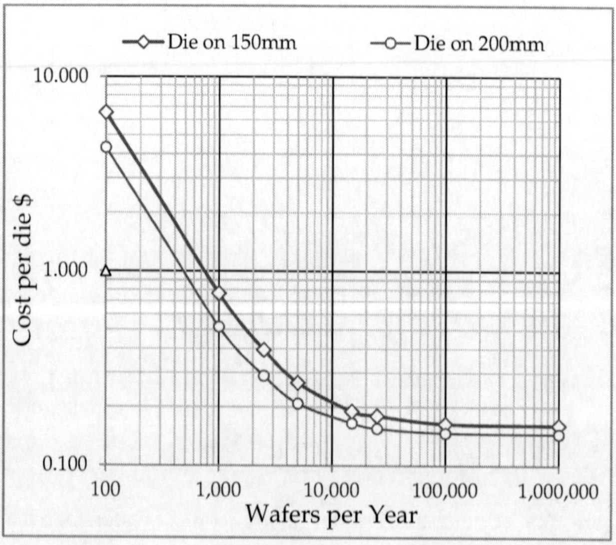

Figure 10.10 Unit cost of a die as a function of wafers per year.

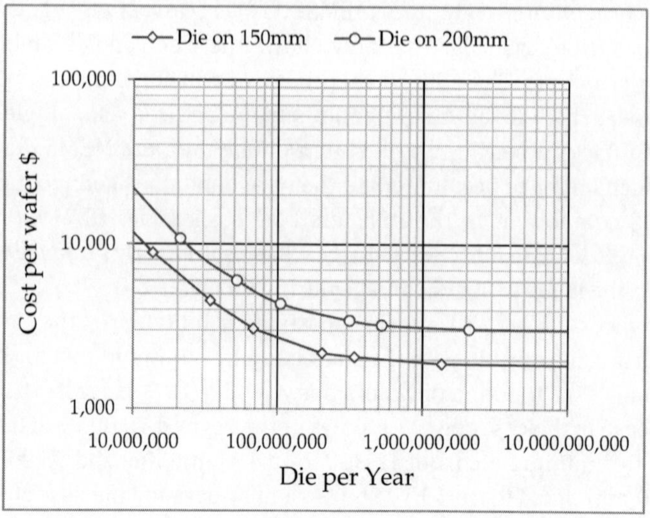

Figure 10.11 Unit cost of a wafer as a function of dies per year.

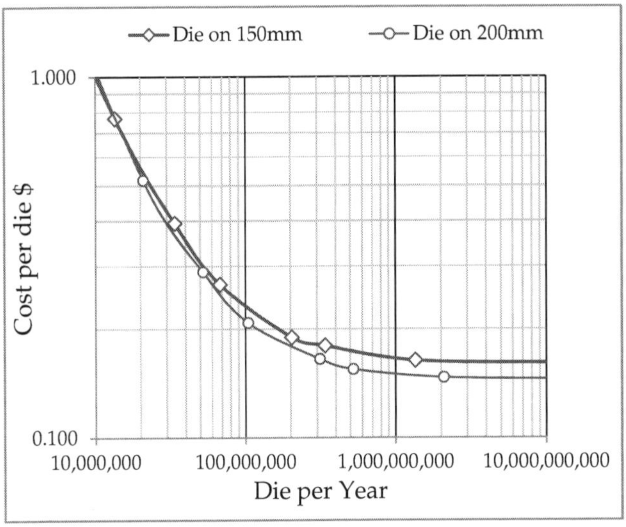

Figure 10.12 Unit cost of a die as a function of dies per year.

Alternatively, the costs may be calculated as a function of the number of dies per year. Figure 10.11 calculates that the decrease in the cost per wafer slows down beyond 200×10^6 dies per year to a minimum cost of $1,860 and $3,000 as before. Figure 10.12 predicts that the cost of dies that are manufactured on a 150 mm line can be reduced by transferring to a 200 mm line when demand exceeds 20×10^6 dies per year.

10.3 Revenue Estimates

The past and future costs of MEMS microphones can be calculated from MEMSCOST and compared with market surveys, in a similar manner to that shown in Chapter 9 for accelerometers.

Table 10.1 shows the estimated die size as a function of the year of introduction. Data in the public domain is sparse, but there is some die size data about Akustica (Johnson, 2011) and Knowles (Yole, 2011). One of the uncertainties with the available data is that Knowles currently dominates with 80% of the microphone market (previously 90%), with Akustica a growing competitor.

Table 10.1 Available data on MEMS microphone dimensions

Year	Item	Manufacturer	Assembly	ASIC mm^2	MEMS mm^2	Total mm^2
2006		Akustica	Single	2.63	0.98	3.61
2006		Knowles	Side		2.56	
2006		ST	Single			2.63
2007		Analog Devices	Single			0.9
2008		Akustica	Single	1.39	0.568	1.96
2008		Knowles	Side		1.82	
2008		ST	Single			2.51
2009		Akustica	Single	0.72	0.28	1
2009		Analog Devices	Single	0.83	1	1.83
2010		Knowles	Side		1.21	
2010		ST	Stack			2
2011		Akustica	Side	0.55	0.16	0.71

Others, such as STMicroelectronics, have entered the market and may significantly change the future market share amongst the companies. Available data on the dimensions of microphones, and also whether they are single- or multi-assembly, is also shown in Table 10.1.

The approach to derive useful data for MEMSCOST calculations has been to estimate an average ASIC size for each supplier, taking into account a 150 mm or 200 mm line and whether the design is single- or multi-die. If multi-die, the ratio between the MEMS microphone dies and their ASIC dies is approximately 1.3:1.

Finally an extrapolation from the known data over the period 2006–2011 can be made for the period 2011–2017, and both sets of data are shown in Fig. 10.13. An important assumption is that while the ASIC may get smaller with decreasing technology nodes, the MEMS component cannot continue to scale down at the same rate. This is illustrated by the projected size of the ASIC and MEMS dies in the near future (by the author).

The most recent market surveys on the volume of MEMS microphone sales are shown in Fig. 10.14 (Bouchaud, 2012; iSuppli, 2010; Yole, 2011). The growth of sales in consumer goods, such as the Apple iPad, has boosted considerably the sales of MEMS microphones. MEMSCOST calculations use the Bouchard data extended from 2015 to 2017 by the author.

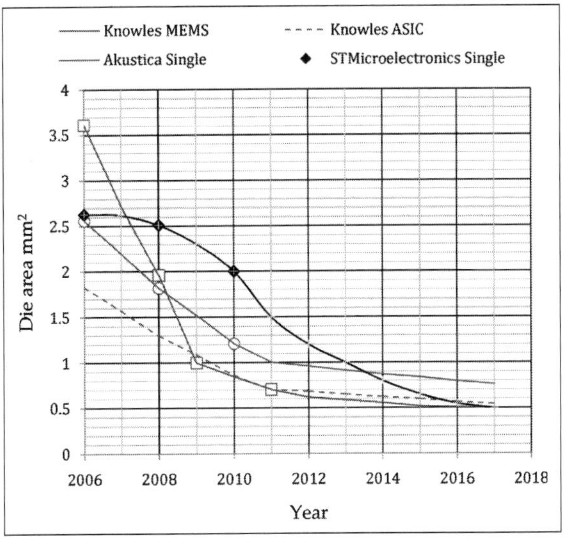

Figure 10.13 Microphone die area (2006–2017). Markers show known dimensions; the remaining dimensions are estimated by the author.

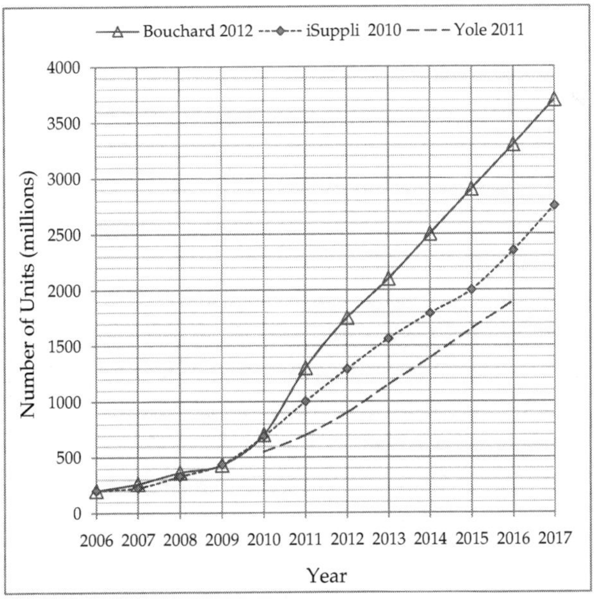

Figure 10.14 Microphones sold per year (2006–2017).

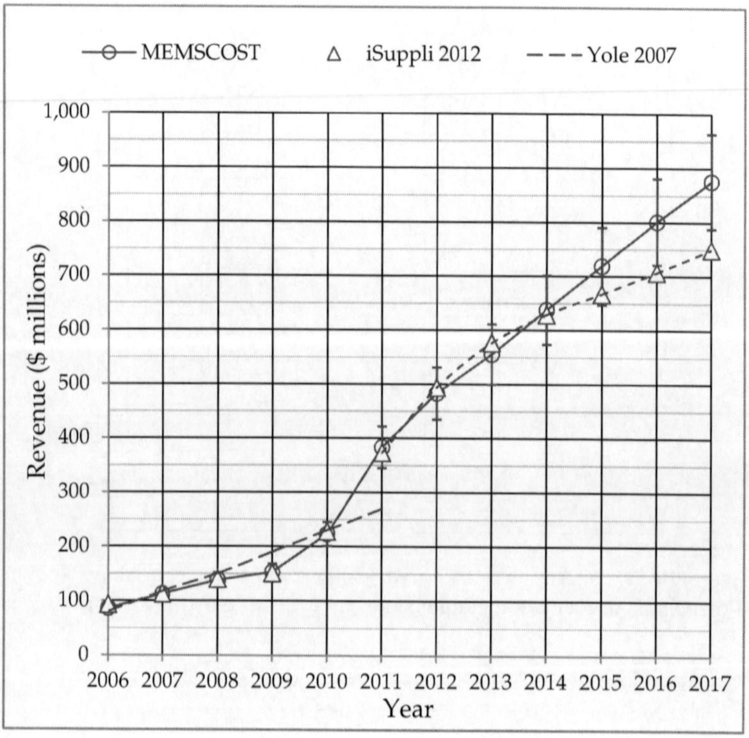

Figure 10.15 Annual revenue from microphone sales (2006–2017).

Market data for revenue from microphone sales is shown in Fig. 10.15 (iSuppli, 2012; Yole, 2007). There has been a considerable increase in sales from 2010 to 2011, as previously noted, and sales by 2015 are projected to reach $667 million per annum. The estimated revenue, as calculated by MEMSCOST and using available data, is a good fit to the market survey data.

Estimates of the market share of 150 mm and 200 mm lines to date are simplified as Knowles (150 mm) has dominated the industry. For future years to 2017, some assumptions must be made as to how this will change, in particular with Analog Devices (150 mm), Akustica (200 mm) and STMicroelectronics (200 mm) manufacturing increasing quantities of microphones. The author's estimate of the evolution of the market share is shown in Fig. 10.16.

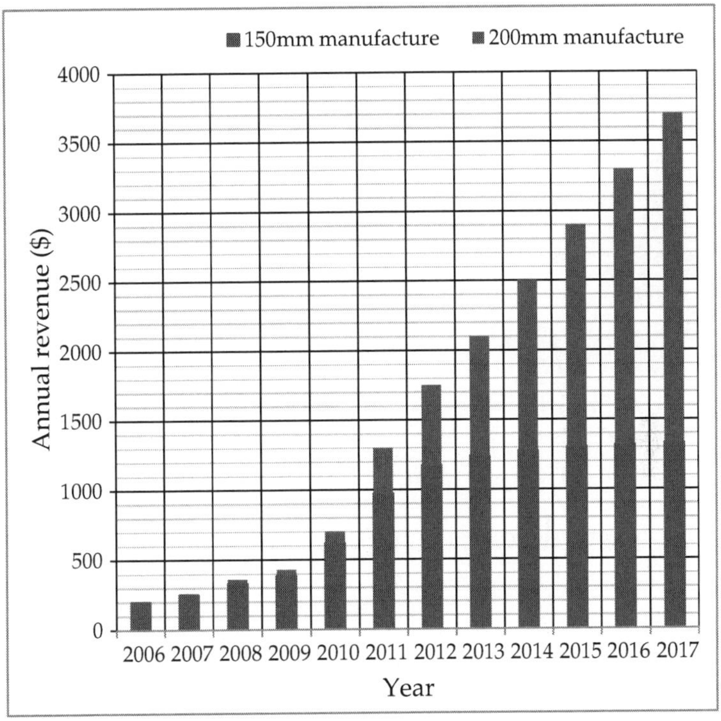

Figure 10.16 Microphone market share between 150 mm and 200 mm lines (author's estimate).

Data from the die size and the number of units sold per year can now be used as input to MEMSCOST to produce an estimate of the average unit cost of a MEMS microphone, in a similar manner to accelerometers in Chapter 9. The results are shown in Fig. 10.17 for tested and packaged devices, assuming an overall yield of 85% (note the 10% error bars). The estimated market cost of a MEMS microphone is obtained by using the number of units from Fig. 10.14, the market share of individual product lines from Fig. 10.15 and the revenue data from Fig. 10.16. Note the cost reductions predicted by the International Technology Roadmap for Semiconductors (ITRS, 2012). From the assumptions used and in comparison with market data, the best fit for the profit margin is 39%.

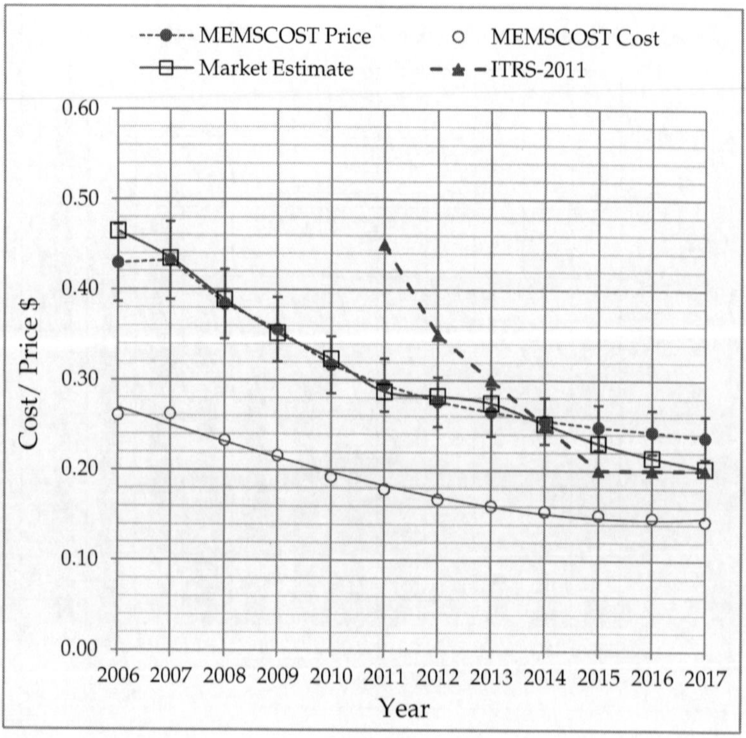

Figure 10.17 Estimated cost of a MEMS microphone (2006–2017) at 85% yield.

10.4 Cost Comparisons for Testing Regimes

An analysis of the comparative costs of different testing regimes, similar to that described for accelerometers in Chapter 9, has been carried out for MEMS microphones. The example chosen is for 500 million, 1 mm × 1 mm single die. The cost of testing at the wafer level, both prior to and after packaging, is compared with the cost of performing a single MEMS test after packaging. The results are similar to those shown in Chapter 9, Fig. 9.21 for accelerometers. In the case of microphone testing, if the yield is above 78%, it is cost-effective to perform only post-packaging testing (Fig. 10.18).

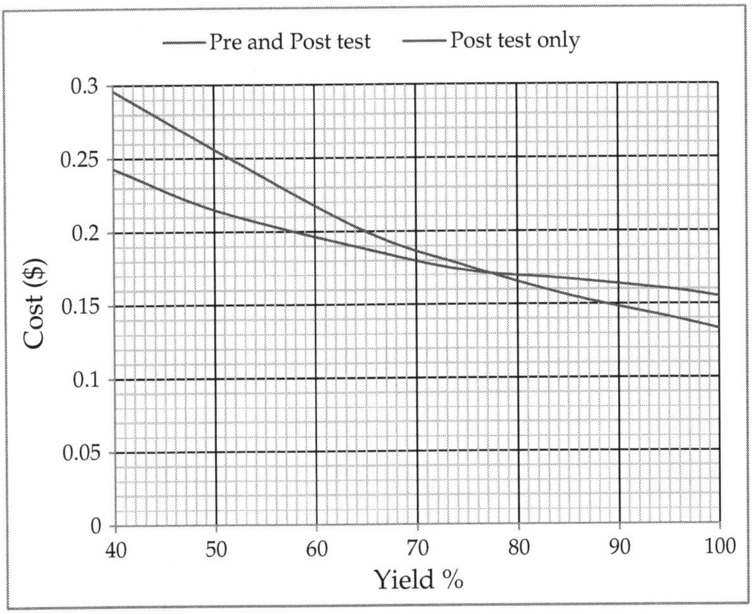

Figure 10.18 Crossover yield at which a change of the testing regime reduces the costs of a microphone (compare with Chapter 9, Fig. 9.21).

10.5 Cost Comparison between Single- and Multi-Die Assembly

Figure 10.19 shows the unit cost of a fabricated, tested and packaged device for the two methods of manufacture (single die and multi-die) and the purchase of the ASIC from a 300 mm pure-play foundry, as with accelerometers in Chapter 9. The example chosen is 500 million, 0.7 mm^2 microphones on a 200 mm line with 85% yield. It can be seen from Fig. 10.19 that the cost advantages of having the ASIC made at a foundry do not completely outweigh the single die's advantages of less silicon and lower packaging costs.

Table 10.2 shows that the cost to an in-house microphone manufacturer with a single die design would be $0.148, whereas a multi-die design would cost $0.160 per microphone. Using an external foundry for the ASIC signal processor, while maintaining construction of the MEMS component in-house, would cost an

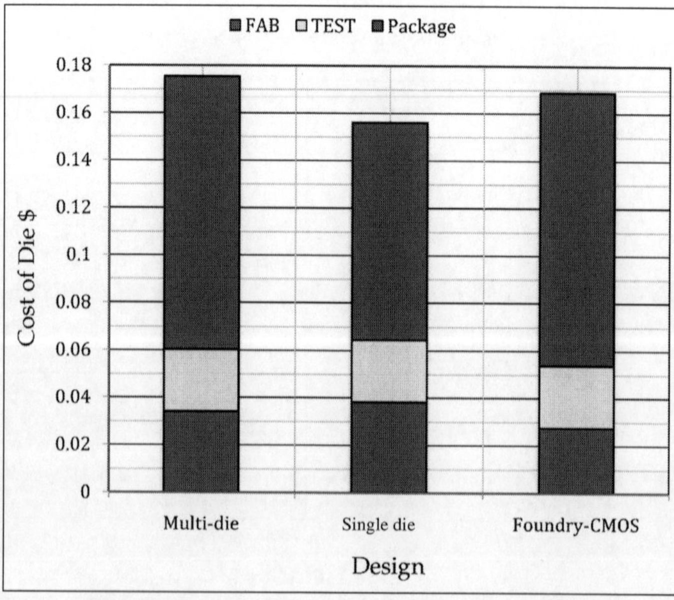

Figure 10.19 Microphones—costs for different manufacturing options.

Table 10.2 Microphones—manufacturing options

	IDM single	IDM multi	ASIC foundry	
ASIC die area	0.7	0.8	0.8	mm^2
ASIC cost	0.0308	0.0172	0.0057	$
Foundry profit margin	–	–	40	%
ASIC cost	0.0308	0.0172	0.0080	$
Fabricate sensor, test and package at IDM				
Sensor (fabrication)	–	0.0035	0.0035	$
Test	0.0263	0.0263	0.0263	$
Package	0.0909	0.1134	0.1134	$
Cost	0.1480	0.1604	0.1512	$

estimated $0.151, a reduction of only 5.7%. This is because the difference in volume between an in-house manufacturer and a foundry does not lead to a large cost differential for the ASIC (see Table 10.2).

10.6 Sensitivity to Manufacturing Costs

MEMSCOST can calculate the sensitivity of the estimated microphone costs to changes in the various financial and manufacturing parameters by using the same equation as for accelerometers, that is, Eq. 9.10 in Chapter 9, Section 9.6. A list of differential cost coefficients and the results of a change of +10% in a given parameter is shown in Table 10.3, compiled in an identical manner to Chapter 9, Table 9.4.

The estimated unit cost of a factory manufacturing 500 × 10^6 microphones per year is \$0.156 for the parameters shown in Table 10.3. Although the microphone and accelerometer coefficients are not the same, the maximum and minimum contributions are similar, that is, yield and the capital cost of the packaging equipment, respectively. This is shown graphically in Fig. 10.20.

The estimated uncertainty in calculating microphone costs can be estimated from the coefficients in Table 10.3 and Eq. 9.11 in Chapter 9, Section 9.6. For a 10% error in all parameters the estimated uncertainty is 14.2%, equivalent to \$0.022, for a 10% change in all parameter costs. If these values were assumed to be

Table 10.3 Differential cost coefficients and % change of overall cost

Parameter	Coefficient		Nominal	10%	Change	
					$	%
Yield	0.0016	$/%	85	8.5	0.00136	8.7
Machines/ technician	0.0535	$/machine technician	2	0.2	0.0107	6.8
Staff hours	4E−05	$/hour	1600	160	0.0064	4.1
Uptime	−0.0006	$/%	90	9	0.0054	3.4
Depreciation	0.005	$/year	5	0.5	0.0025	1.6
Wafers/year	4E-11	$/wafer	1.00E+08	1.00E+07	0.0004	0.2
Capital package	−8.0E-05	$/%	100	10	−0.0001	−0.1
Maintenance	−0.0015	$/%	2	0.2	−0.0003	−0.2
Capital test	−7E-05	$/%	100	10	−0.0008	−0.5
Technology node	−0.0015	Per node	0.18	1	−0.0015	−1
Capital fab.	0.00018	$/%	100	10	−0.0018	−1.2
Die area	−0.024	$/mm^2	1	0.1	−0.0002	−1.5
Salary	−1E-6	$/k$	47083	4708	−0.0040	−2.6
Overheads	0.00064	$/%	150	15	−0.0096	−6.1

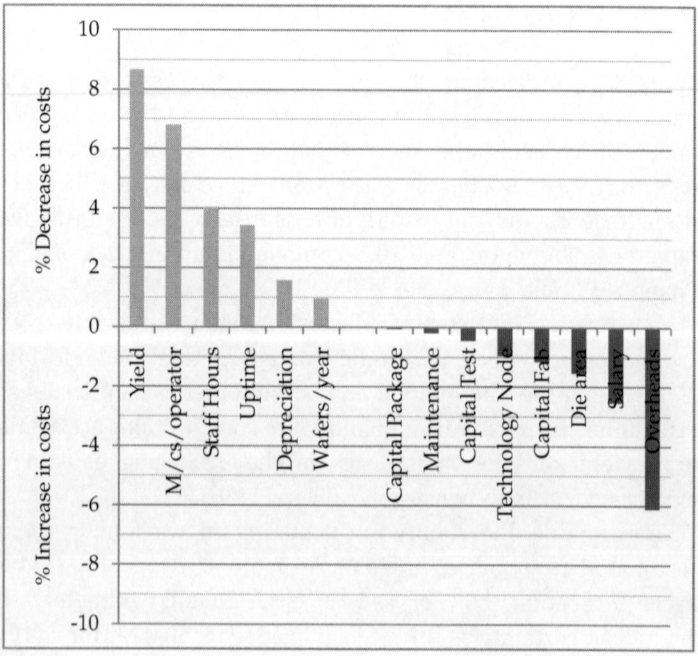

Figure 10.20 Cost sensitivities—% change in overall costs per parameter.

1 standard deviation, then the unit costs would lie between \$0.112 and \$0.190 (2 standard deviations).

References

Dixon-Warren, St, J (2009). Analog Devices wins microphone socket in the New iPOD. *Mironews,* December(86).

Yole (2011). MEMS microphones. *Yole Report 2011.*

Johnson, R (2011). World's smallest MEMS microphone. www.eetimes.com

Bouchaud, J (2012). Booming iPad sales help Apple to become largest consumer of microphones in 2011. www.isuppli.com/MEMS-and-Sensors.

Bouchaud, J (2010). MEMS microphones gain volume in 2010 and set to make more noise. *Microphone Special Report.* www.isuppli.com/MEMS.

Dehe, A (2006). Silicon microphone development and application. *Sens. Actuators A*, **133**, 283–287.

Feiertag, G, Winter, M, and Leidl, A (2009). Packaging of MEMS microphones. *Smart Sens., Actuators MEMS, Proc. SPIE*, **7362**.

ITRS (2011). Micro-electro-mechanical systems. *International Roadmap for Semiconductors. 2011 Edition Micro-Electro-Mechanical Systems.*

Schomburg, WK (2011). *Introduction to Microsystem Design.* Springer-Verlag, Berlin.

Chapter 11

MEMS Foundries

11.1 Introduction

Business models for integrated circuit (IC) and microelectromechanical systems (MEMS) manufacture can be divided into in-house and foundry operations. Companies operating the in-house model have both design and manufacturing capabilities and produce their own products (e.g., Intel, Samsung, STMicroelectronics, Bosch). A foundry offers a manufacturing service to other companies, using state-of-the-art technology nodes in the case of IC foundries and a variety of different technologies in the case of MEMS foundries (e.g., bulk or surface microengineering). Foundries offering a manufacturing service only are known as 'pure play'. Companies without their own manufacturing capability are known as 'fabless'.

Foundries started to develop in the semiconductor industry as the cost of constructing in-house facilities reached billions of dollars (see Rock's law, Chapter 7, Fig. 7.6), and this tendency accelerated when the cost of periodically replacing equipment for a new technology node reached similar levels. Foundries such as the United Microelectronics Corporation (UMC), founded in 1980, the Taiwan Semiconductor Manufacturing Company (TSMC), founded in

MEMS Cost Analysis: From Laboratory to Industry
Ron Lawes
Copyright © 2014 Pan Stanford Publishing Pte. Ltd.
ISBN 978-981-4411-06-6 (Hardcover), 978-981-4411-07-3 (eBook)
www.panstanford.com

1987 and Global Foundries founded in 2009 entered the field and have since grown to become billion dollar 'pure play' enterprises.

Over the last 20 years, many MEMS foundries have been set up, usually offering a limited range of manufacturing processes so that, for example, one foundry will offer microfluidic processes and another bulk micromachining. The MEMS industry has, in part, followed the semiconductor business model by using specialist MEMS foundries, rather than developing in-house facilities, although the major players still retain in-house facilities. With the development of MEMS technology based on silicon and the growing size of the market, even big players in complementary-metal-oxide semiconductor (CMOS) fabrication, such as the TSMC, now offer a MEMS fabrication service.

11.2 MEMS Foundries

Business models for a MEMS foundry vary from smaller units, undertaking prototyping and small/medium production runs, to larger units moving towards just a few processes and possibly in-house products. A potential future trend is from in-house production to the use of a foundry service, even if it means changing the design strategy. For example, Analog Devices started its accelerometer series in 1991 with its in-house iMEMS factory producing a single die design. By 2009 its strategy had changed to a multi-die design, with the application-specific integrated circuit (ASIC) die being supplied by a large-scale foundry (TSMC) enabling a more advanced, cost-effective processing capability (Dixon-Warren, 2009).

There are significant differences between an IC foundry and a MEMS foundry, both in scale and the type of product manufactured. MEMS foundries also differ in that an IDM MEMS foundry (e.g., Bosch, STMicroelectronics) will have a capacity for manufacturing both the signal-processing ASIC and the MEMS sensor/actuator in the same silicon process, while a smaller-scale MEMS foundry (e.g., Tronics, Micralyne) will only manufacture the MEMS components.

Revenue for a MEMS foundry is an order of magnitude lower than for an IDM in large-volume production. Revenues for 2011 (Yole, 2012) are shown in Fig. 11.1.

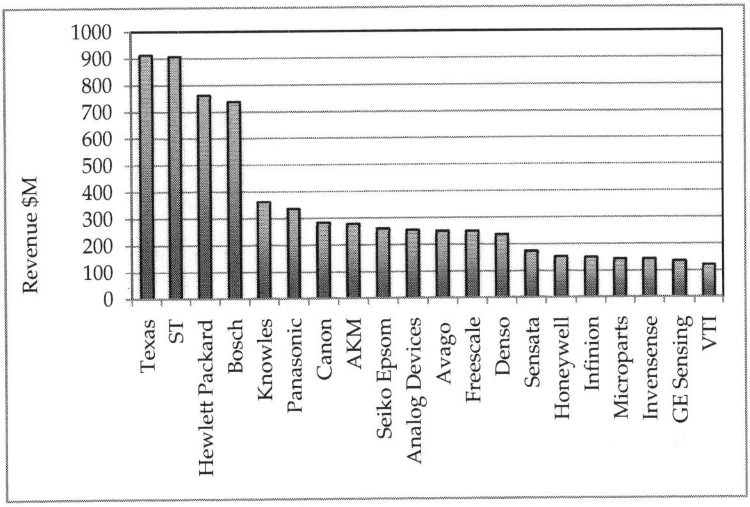

Figure 11.1 MEMS foundry revenue for 2011 (Yole, 2012).

11.3 Multi-Product Use of Processes and Equipment

Cost calculations for a multi-product MEMS foundry are more complicated than for a single-product IDM factory, as the overheads and equipment can be shared across the various customers. MEMSCOST has a 'foundry' mode that takes into account all jobs completed on an annual basis.

For each job, the MEMSCOST *Process Library* enters the machines used and the time, materials and all resources needed to process each task. This data is loaded sequentially into the *Foundry Cost Simulator*, which then calculates the total number of machines, the total machine time and all the resources used. A standard fixed plus variable cost calculation is then performed to produce an estimate of the annual charges per machine hour.

For example, an in-house MEMS foundry set up to manufacture 5,000 wafers per year containing a 1 mm × 1 mm accelerometer ASIC die (over 100 million dies) would fabricate, test and package each die for a cost of $0.211/device. A large-scale foundry, set up to manufacture 25,000 wafers per year for a number of customers, would manufacture the same ASIC die for $0.157, a cost reduction of

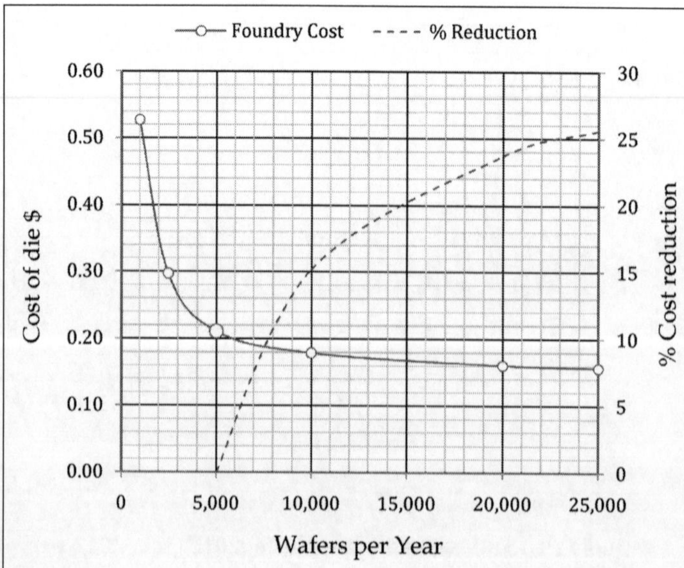

Figure 11.2 Cost benefits moving from in-house to a large-scale foundry.

25.6%. The cost and cost reduction, as functions of wafer output, are illustrated in Fig. 11.2.

11.4 MEMS Foundry Lightly Loaded

A problem for a new MEMS foundry is the adverse effect of building and operating a foundry that is subsequently not fully loaded. MEMSCOST enables some basic data to be calculated. Figure 11.3 illustrates the effect of loading on a foundry designed to produce 200 million 1 mm × 1 mm accelerometers from 10,000 200 mm diameter wafers, where the cost of production (in millions of dollars) is given by

$$\text{Cost} = 2.874/10^3 N + 7.97 \tag{11.1}$$

Assume that the MEMS foundry has been constructed to provide 10,000 wafers a year, with a mix of jobs, and that the business plan calls for the foundry to have a 40% profit margin when loaded

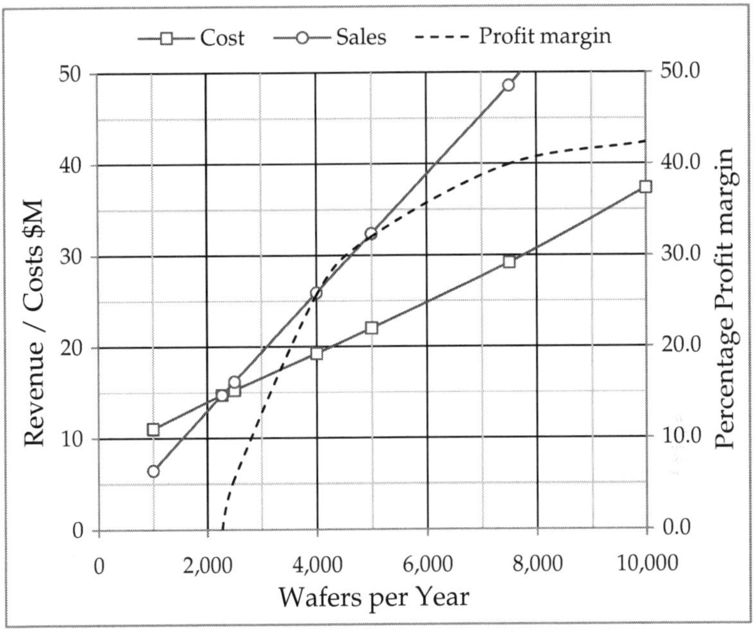

Figure 11.3 Effect of production loading on cost and revenue.

at 75% of full capacity. The cost of operating at full capacity is calculated by MEMSCOST at $37.3 million.

As the demand for accelerometers falls below the planned value, some costs (e.g., the depreciation of equipment) remain fixed but manufacturing costs reduce with declining production. Assuming the sale price is maintained, as the sales decrease there comes a critical point at which the revenue does not cover the total cost of manufacture. In this example, by reducing the manpower levels according to demand (where possible), the foundry would be profitable until the sales fell to 30% of the designed output.

Table 11.1 shows what happens to the profit margin if the calculated equipment and process charges are defined and held constant throughout the trading year as volumes increase or decline. It will be seen that at 2,773 wafers per year or 30% of expected loading (2,273/7,500), the foundry becomes unprofitable. More business for the foundry must be found or the price increased. Alternatively, if trading exceeds expectations by 33%, that is, to

Table 11.1 Notional annual output for a MEMS foundry

Wafers	Cost		Revenue		Profit
	$/wafer	$	$/wafer	$	%
10,000	3,734	37,342,893	6,475	64,750,000	42.3
7,500	3,885	29,173,916	6,475	48,562,500	40.0
5,000	4,401	22,006,043	6,475	32,375,000	32.0
4,000	4,816	19,262,098	6,475	25,900,000	25.6
2,500	6,110	15,275,800	6,475	16,187,500	5.6
2,273	6,473	14,714,110	6,475	14,717,675	0

the full foundry capacity of 10,000 wafers per year, then the profit margin increases to 42.3%.

11.5 Multi-Project Wafers

In the early 1980s, there was a demand for semiconductor processes from the academic community, but the cost of fabrication was prohibitive, even if access to a foundry was possible. When foundry manufacture of ASICs developed, as opposed to in-house IDM, the concept of the multi-project wafer (MPW) evolved. The basic concept of an MPW was to share the increasingly dominant cost of the mask set amongst many users and to identify foundries that were prepared to produce an amalgamated mask and subsequently die/wafers for each individual user. Parallel developments in software have produced standard design rules in the form of design kits for the potential user.

The MPW concept for ICs is now well established through organisations such as Europractice, supported by the European Union, and by MOSIS in the United States. Major IC manufacturers offer an MPW project service through such organisations, as well as providing their own service. The MPW concept was not immediately applicable to MEMS as the manufacturing processes were anything but standard (unlike IC semiconductor processes) and software support was lacking. The status of standardised silicon MEMS processes has much improved over the last decade, and there are now several standard MEMS surface-micromachined processes, notably the Sandia ultra-planar, multi-level MEMS technology (SUMMIT) V process from Sandia Laboratories and various multi-user MEMS processes (MUMPS) from MEMSCAP (Hardy, 2007) and Circuits Multi-Projects (CMP) (DiPendina, 2011). Software design kits are available for these processes (Kubby, 2011).

PolyMUMPS is a MEMSCAP standard silicon surface micro-engineered process. MetalMUMPS has electroplated nickel as the primary structural material, with doped polysilicon for additional structure, silicon nitride as an electrical isolation layer and silicon oxide as the sacrificial layer. SOIMUMPS has silicon-on-insulator (SOI) as the starting substrate so that the silicon can be patterned down to the oxide layer. Both MetalMUMPS and SOIMUMPS processes are relatively simple but can manufacture novel devices.

SUMMiT V and PolyMUMPS standardise the key process steps of polysilicon deposition for structural materials and the etching of sacrificial SiO_2 to release structures. Thousands of designs have now been manufactured for both academia and industry using these processes. The entry cost of MEMS fabrication to access a mainstream foundry (if access is at all possible) would be hundreds of thousands of dollars. The MPW concept has brought the research and development (R&D) costs down to $5-$17,000 for a few samples.

MEMSCOST can be used to estimate the cost of these processes across a range of scenarios and a variety of assumptions. These estimates are based on the commercial cost of reticles/masks, the in-house fabrication cost of the number of wafers to be processed, a charge per user to incorporate each device into the mask/reticle design and a charge to run the service.

The internal costs of the various organisations offering an MPW service are not publically known, so the MEMSCOST calculations are based upon typical cost estimates made by the author. Similarly, the annual output for a given foundry is unknown, although it is commonly believed that commercial MEMS foundries produce between a few thousand to a few tens of thousands of wafers per year. Technical descriptions for the SUMMiT V (Sandia, 2008) and PolyMUMPS (Carter, 2005) processes are publically available.

Sandia's Agile SUMMIiT V process offers a five-layer, polysilicon process with four layers of sacrificial oxide and one metal layer, as

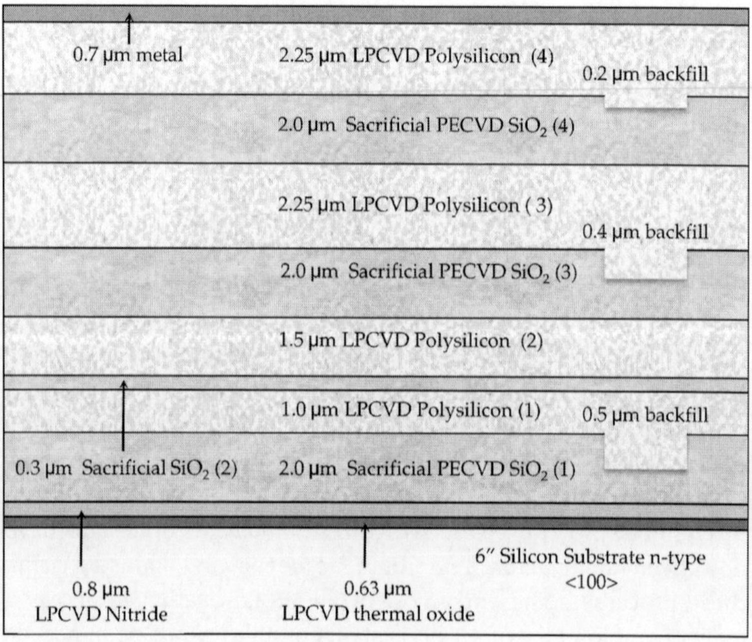

Figure 11.4 Schematic of the SUMMiT V process.

shown in Fig. 11.4. The process is assumed to use a wafer stepper to meet the 0.5 μm minimum feature size and different MEMS designs are assembled onto a 14-reticle mask set (see Fig. 11.5). A die size of 16.5 mm × 12.7 mm is fixed so that eight 6.34 mm × 2.82 mm modules can be manufactured per die. Sandia uses a 150 mm diameter wafer, so 63 dies can be fitted per wafer (500 modules), and as each customer receives 200 unreleased modules, a single wafer can provide the modules for at least two customers. Assuming the stepper has a reticle demagnification of ×5, it is possible to fit a 2 × 2 array into a single reticle so that four customers can share the cost of the reticles. An MPW service offers a fixed starting date and a fixed number of runs per year, for example, Sandia offers 10 runs per year. The pricing for such an MPW depends not only on the internal costs for wafer fabrication but also strongly on the cost of reticles that are presumed to be purchased from an external commercial mask shop.

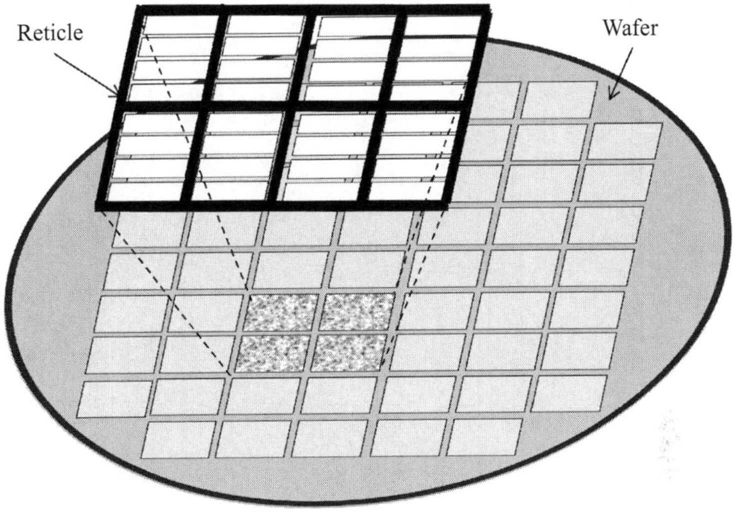

Figure 11.5 Schematic of an MPW wafer using a 2 × 2–die reticle.

Table 11.2 shows a possible range of costs for purchasing the reticles (a significant fraction of the overall cost) and fabricating the wafers and an internal administrative cost for preparing the MPW reticles.

Table 11.2 Component costs of the Sandia SUMMiT V process (author's estimates)

	Reticles		Wafers				
Users/run	Sets	Costs	No.	Costs	CAD	Total	Cost/user
1	1	35,000	1	450	100	35,551	35,551
2	1	35,000	1	450	200	35,651	17,826
3	2	70,000	2	900	300	71,202	23,734
4	2	70,000	2	900	400	71,302	17,826
5	2	70,000	3	1,350	500	71,853	14,371
6	3	105,000	3	1,350	600	106,953	17,826
7	3	105,000	3	1,350	700	107,053	15,293
8	4	140,000	4	1,800	800	142,604	17,826
9	4	140,000	4	1,800	900	142,704	15,856
10	4	140,000	5	2,250	1,000	143,255	14,326

Abbreviation: CAD, computer-aided design.

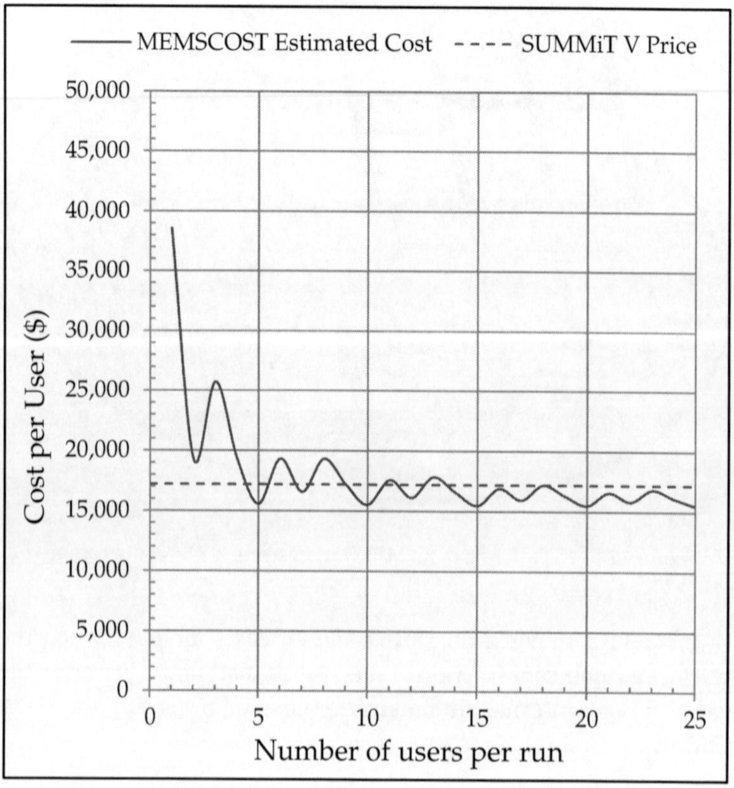

Figure 11.6 Estimated cost for the SUMMiT V process as a function of the number of users per fabrication run.

Sandia provides 200 modules to the user at a 2011 price of \$17,200 (Sandia, 2011) (\$480/cm^2). This should cover the actual cost of performing the work but not the cost of R&D programmes that produced the technology. This price is shown in Fig. 11.6 in comparison with the MEMSCOST estimated cost. If the number of users for a specific run increases beyond the five user designs that can be fitted, then reticle sets will have to be designed and purchased and another wafer processed to meet the delivery quantities of modules. A minimum number of five users per run are required to cover approximately the cost of manufacture. This analysis suggests that the price charged is close to the cost of manufacture, depending on the cost of purchasing masks.

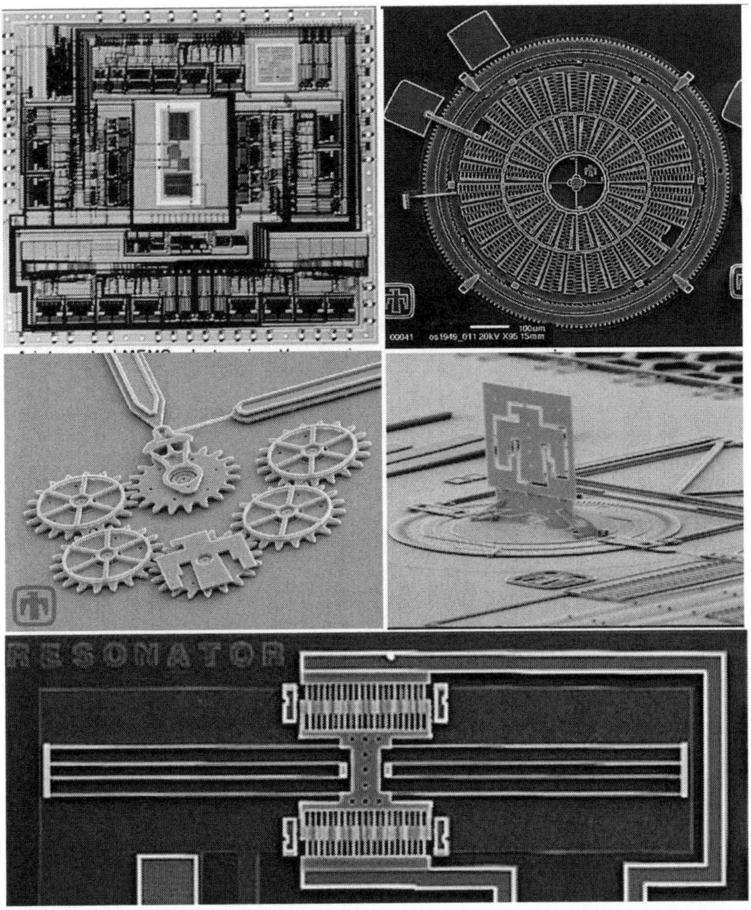

Figure 11.7 SUMMIT™ Technologies showing an integrated three-axis accelerometer (top left), a torsional ratching actuator (top right), a vertical mirror on a rotary-indexing stage (middle right) and a six-gear assembly capable of 250,000 rpm (middle left). The bottom figure shows a comb-drive resonator (courtesy of Sandia National Laboratories).

Examples of the complex devices that can be realised with Sandia's 'Samples' design suite, SUMMIT V and microelectronics capabilities are shown in Fig. 11.7. The examples are taken from Sandia's web page of Integrated μ ElectroMechanical Technologies and Electrostatic Actuators (Sandia, 2005–2008).

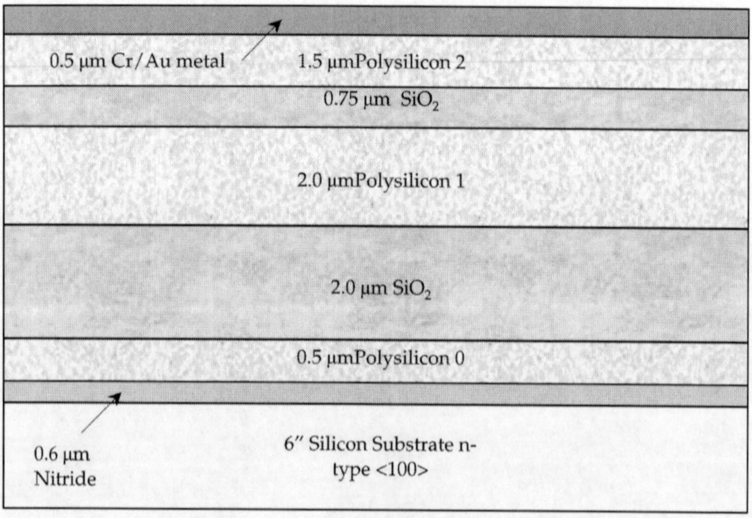

Figure 11.8 Schematic of the PolyMUMPS process.

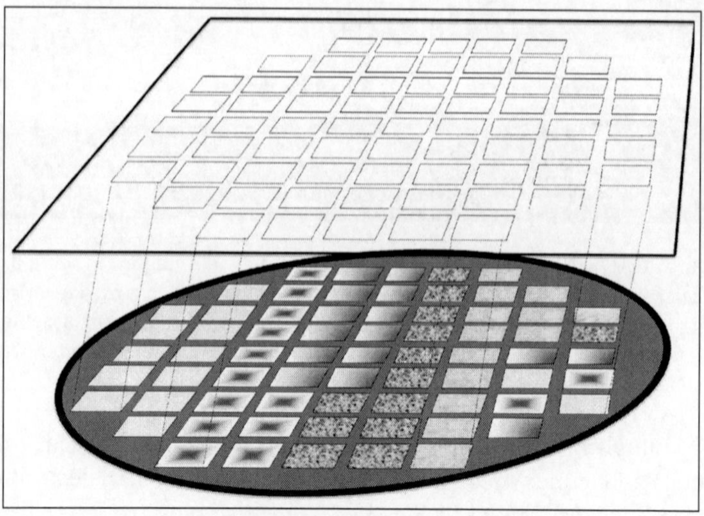

Figure 11.9 Conceptual layout of an MPW wafer, using a 1:1 mask and a wafer aligner.

PolyMUMPS offers three layers of polysilicon with two layers of sacrificial oxide and one metal layer (see Fig. 11.8). The PolyMUMPS process only requires lithography resolution to 2 μm; hence the use of a wafer aligner will be assumed. The different MEMS designs are assembled into a single mask set, that is, one mask per wafer layer, as shown in Fig. 11.9. PolyMUMPS uses a die size of 10 mm × 10 mm on a 150 mm wafer and can fit 132 dies into a wafer. This allows the output for eight users to be fitted on to the mask. MEMSCAP provides 15 dies to each user.

Table 11.3 shows the estimated costs, calculated by MEMSCOST, as a function of the number of users per run (MEMSCAP typically offers 10 runs per year). It is assumed that some R&D costs would be recovered.

The 2011 prices charged by MEMSCAP (MEMSCAP, 2011), CMP (CMP, 2011) and Europractice (Europractice, 2012) are shown in Fig. 11.10, although the cost structures for PolyMUMPS, from MEMSCAP, CMP and Europractice, are not in the public domain. The MEMSCAP price of \$5,300 (\$350/cm^2) is the normal charge to the industry for the service. However, MEMSCAP and Europractice charge academia a discounted price of \$3,700 and \$3,500 (\$233/cm^2), respectively (2011 prices). This discounted price is the nearest indication available for the actual foundry fabrication costs.

Table 11.3 Component costs of the MUMPS process (author's estimates)

	Reticles		Wafers				
Users/run	Sets	Costs	No.	Costs	CAD	Total	Cost/user
1	1	12,952	32	11,200	100	24,248	24,248
2	1	12,952	32	11,200	200	24,384	12,192
3	1	12,952	32	11,200	300	24,484	8,161
4	1	12,952	32	11,200	400	24,584	6,146
5	1	12,952	32	11,200	500	24,684	4,937
6	1	12,952	32	11,200	600	24,784	4,131
7	1	13,216	32	11,200	700	24,884	3,555
8	1	13,216	32	11,200	800	24,948	3,123
9	2	25,900	32	11,200	900	38,036	4,226
10	2	25,900	32	11,200	1,000	38,136	3,814

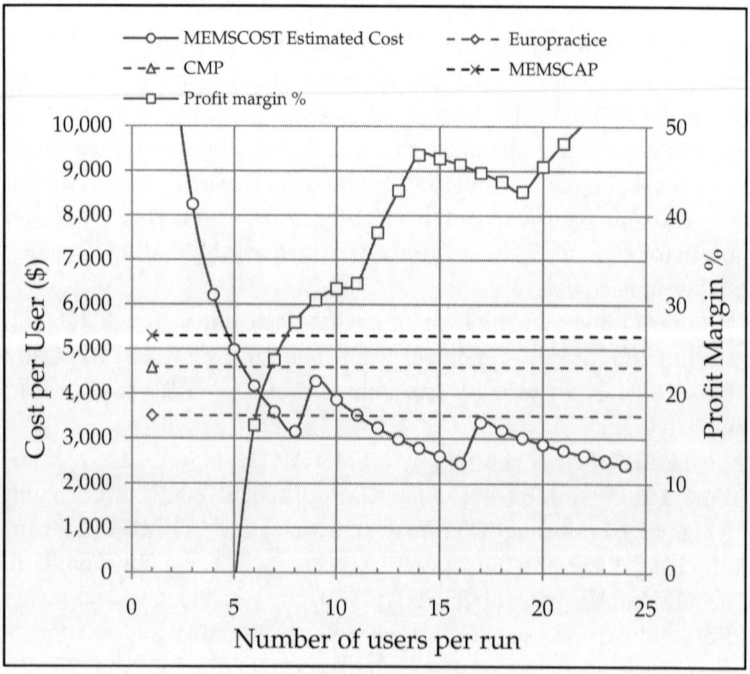

Figure 11.10 Estimated cost for the PolyMUMPS process as a function of the number of industrial users per fabrication run.

The MEMSCAP and Europractice prices fit well into the estimates made in Table 11.3.

Also shown in Fig. 11.10 is the estimated profit margin from the turnover from the industry. Note that the number of users for each run is variable, which will significantly affect mask costs. The problem for the foundry is to estimate the likely number of users per run and hence the annual price. Figure 11.10 suggests that, on average, the minimum number of users must be equal or greater than five per run. An average profit margin of 32% will be produced if the number of users per run increases to 6–11.

References

Carter, J, et al. (2005). *PolyMUMPS Design Handbook,* Revision 11.0. MEMSCAP Inc.

CMP (2011). Multi project circuits. cmp.imag.fr/products/ic/?p=prices

DiPendina, G (2011). MEMS processes from CMP. *CMP Seminar Grenoble.*

Dixon-Warren, St, J (2009). Chipworks: inside Analog devices' new MEMS strategy. http://memsblog.wordpress.com/2009/09/28/chipworks

Europractice (2012). Europractice MEMS MPW runs price list. europractice-ic.com/MEMS_pricing.php

Hardy, B (2007). Introduction to MEMSCAP and MUMPS. *Europractice STIMESI MEMS and System in Package Workshop IMEC.*

Kubby, JA (2011). *A Guide to Hands-On MEMS Design and Prototyping.* Cambridge University Press. 7–13.

MEMSCAP (2011). Pricing. memscap.com/products/mumps/soimumps/pricing

Sandia (2005–2008). Capabilities—actuators. mems.sandia.gov/about/ actuators.html

Sandia (2005–2008). Capabilities integrated uElectroMechanical technologies. mems.sandia.gov/about/electro-mechanical.html

Sandia (2011). Samples program price list. mems.sandia.gov/samples/pricing

Sandia (2008). SUMMiT V five level surface micromachining technology design manual. *Sandia SAND number: 2008-0659P.*

Yole (2012). Yole development top 30 MEMS ranking. www.yole.fr/2012_press release.aspx

Chapter 12

Financial Reporting and Analysis

12.1 Introduction

MEMSCOST is designed to compute the cost structure of a device from an engineering perspective. However, this same data can also provide much of the raw data for simple accounting statements and financial metrics, and MEMSCOST itself converts the engineering cost data into a simple profit & loss statement. These same data options can be used as the basis of an investment business case. This chapter will summarise the role of the key financial statements (it is not intended as a lesson in accounting) and show how MEMSCOST can be used in financial assessments by presenting a simple, illustrative investment business case. Finally, this chapter will look at some key financial concepts.

12.2 Key Financial Statements

Many countries have their own accounting standards or 'generally accepted accounting practices/principles' (GAAP), for example, US GAAP and standards issued by the International Accounting Standards Committee (IASC), known as International Accounting

MEMS Cost Analysis: From Laboratory to Industry
Ron Lawes
Copyright © 2014 Pan Stanford Publishing Pte. Ltd.
ISBN 978-981-4411-06-6 (Hardcover), 978-981-4411-07-3 (eBook)
www.panstanford.com

Standards (IAS). Since 2001 the International Accounting Standards Board (which replaced the IASC) has been working on developing globally acceptable accounting standards known as International Financial Reporting Standards (IFRS, 2012), which have been adopted by European Union countries, among others.

Key financial statements using IFRS terminology include:

- the statement of financial position, also known as the balance sheet. This is a 'snapshot' of the business assets and liabilities at a point in time. Assets are balanced by liabilities and equity.
- the statement of comprehensive income. This is a statement of the financial performance over a period of time, showing the income and related costs and outgoings of a business over that period. It includes:
    - the income statement, also known as profit & loss or revenue statement; and
    - the statement of other comprehensive income (which includes gains and losses not shown in the profit & loss, such as derivatives, actuarial gains and sale of securities).
- the statement of changes in equity. This shows the changes in the equity during a period, where retained profits increase and business losses reduce the value of the equity. Investing funds into a business also increases the equity, while a repayment of capital or payment of dividends reduces the equity.
- the statement of cash flows. This reflects a company's liquidity by showing the flow of cash and cash equivalents (safe, readily convertible assets such as short-term government bonds) into and out of a business.
- explanatory notes to the accounts.

12.3 Business Case

This simple business case consists of a series of assumed inputs relating to the investment itself, planned production, revenue and costs over several years. These inputs are modelled in MEMSCOST. As this business case is for illustrative purposes it will be evaluated simply through its profitability and cash flow.

12.3.1 *Product Specification*

Company X wants to build a new factory to produce approximately 500 million MEMS devices, for example, accelerometers. These devices will be produced as 1 mm × 1 mm die on a 200 mm wafer line, with 0.18 um application-specific integrated circuit (ASIC) technology requiring 25,000 wafers per year. The costs associated with this specification are calculated in MEMSCOST.

For simplicity, it is assumed that Company X is able to fund this investment from its own bank balance, although other means of funding are possible. For example, it is also possible to borrow cash externally and pay financing charges.

12.3.2 *Investment*

Figure 12.1 shows a simple timeline for the construction and commissioning of the MEMS factory/foundry. Assuming some preliminary work has been done, a building similar to that shown in Chapter 7 could be constructed within six months. A further three-month period would see the clean rooms and plant being constructed and commissioned. The clean rooms would not be

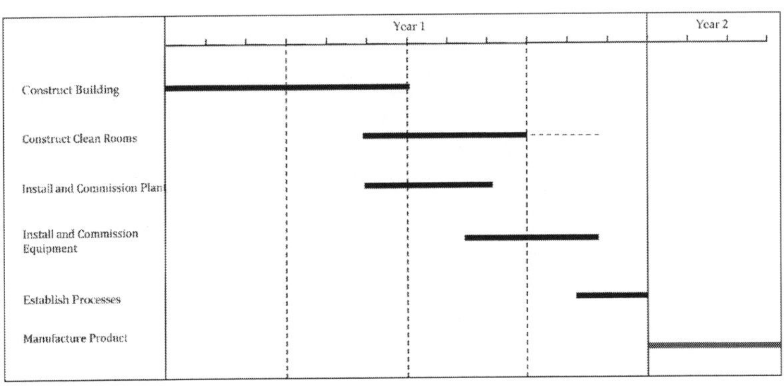

Figure 12.1 Timeline for model factory/foundry build.

brought up to full operation until all equipment had been positioned and all 'dirty' assembly work completed.

In quarter 4 equipment is being fully commissioned and used to establish the processes required. This necessitates the technicians and operators to be in post and therefore the cost of salaries, overheads and materials is included. Also, the initial routine maintenance schedules will be started. It is assumed that the manufacturing process had been developed off-line and was imported into the new factory/foundry.

12.3.3 *Capital Investment*

The initial investment in capital of $127.15 million is shown in Table 12.1, which also shows the period in years over which each capital class is depreciated and therefore the annual charge to the profit & loss account. It is assumed that the purchase of the buildings, clean rooms, plant and manufacturing equipment is paid by cash and the whole investment takes place in year 1. The factory is ready for production on the first day of year 2, and the capital assets are depreciated from that point. There is a fuller explanation of depreciation in Section 12.4.

12.3.4 *Start-Up Expenses*

The start-up expenses in year 1 are the cost of labour and materials during commissioning, maintenance and administration costs, totalling $21.3 million, as shown in Table 12.2.

Table 12.1 Capital investment and annual depreciation

Capital	Investment ($)	No of years	Depreciation
Equipment	67,875,000	5	13,575
Clean rooms	28,035,309	10	2,803,531
Plant	14,750,000	15	983,333
Buildings	16,485,678	25	659,427
Total	127,145,987		18,021,291

Table 12.2 Costs incurred bringing the factory to readiness in year 1

Expenditure	$ (million)
Direct labour	3,516427
Materials	12,289,952
Maintenance	251,496
Administration costs	5,274,641
Total	21,332,516

12.3.5 *Production*

It is assumed that in year 2 only 10,000 wafers will be produced, in year 3 20,000 and from year 4 the planned full load of 25,000 wafers per year (over 500 million devices) will be achieved. It is assumed that in year 2 and year 3 labour and maintenance will be at the full annual rate but that materials will be pro rata with the output. Table 12.3 shows the revenue and costs associated with this production plan, along with the market price per device, assuming a 40% profit margin is maintained.

Table 12.3 Revenue and costs according to MEMSCOST

	Year 1	Year 2	Year 3	Year 4	Year 5
Number of wafers	-	10,000	20,000	25,000	25,000
Number of devices[1]	-	209×10^6	417.9×10^6	522.4×10^6	522.4×10^6
Price per device	-	0.298	0.269	0.262	0.262
Revenue	-	62,238,155	112,214,722	136,706,697	136,706,697
Cost of sales					
Direct labour	3,516,427	10,549281	10,549281	10,549281	10,549281
Materials	12,289,952	14,747,943	29,495,886	36,869,857	36,869,857
Maintenance	251,496	754,489	754,489	754,489	754,489
Depreciation[2]	-	18,021,291	18,021,291	18,021,291	18,021,291
Commercial, general & administration	5,274,641	15,823,922	15,823,922	15,823,922	15,823,922
R&D	-				
Revenue–Costs[3]	−21,332,516	20,362,520	55,591,144	72,709,148	72,709,148

[1]Number of devices per wafer (24,584) at 85% overall yield.
[2]Depreciation of equipment, clean rooms, plant and factory buildings.
[3]This is effectively the profit before taxation (PBT) (see Table 12.4 for a worked example).
Abbreviation: R&D, research and development.

Table 12.4 Income statement $ (profit & loss account)

Revenue	136,706,697
Cost of sales	
Total cost of sales	66,194,920
Gross profit	70,511,777
Commercial, general & administration	15,823,922
R&D	0
Operating profit	54,687,855
Other income and expenses	0
PBT	54,687,855
Tax at 30%	21,812,744
PAT[1]	32,875,111

[1]For tax purposes, it is assumed that depreciation is not an eligible expense and therefore should be added back to the PBT before applying tax, giving a taxable profit of $72,709,148. A 30% tax rate is assumed.
Abbreviation: PAT, profit after tax.

The data shown in Table 12.3 can be used to generate an income statement (profit & loss account). Table 12.4 shows this income statement (Investopedia, 2011) for the first full year of production (year 4).

Simple definitions of the key lines of the income statement are shown below:

- Revenue: Income/sales from products or services sold in the period.
- Cost of sales: The costs associated with revenue generated. This includes both direct costs (direct labour, such as a machine operative, and direct materials, such as the wafer) and factory overheads such as manufacturing management, maintenance of the machines. MEMSCOST includes the depreciation of production assets.
- Gross profit: Revenue less cost of sales.
- Commercial (selling), general & administration: Expenses incurred as a result of supporting the business, for example, expenses to sell the product (sales force, marketing costs) and running or managing the business, such as management, finance and human resources (HR) employees, rent and office supplies.

- R&D: Research costs are incurred in gaining new scientific or technological knowledge and written off as incurred. Development refers to creating a new product or significantly improving an existing one. If these costs are appropriate and can be clearly captured, they can be capitalised. The business case assumes no R&D is required
- Operating profit: Profit arising from the ongoing normal operations of the business.
- Other income/expense: Income and expenses from non-operating sources. Examples include profits/losses from the sale of assets or disposal of businesses or income from investments.
- PBT: Profit Before Tax. Total income less total expenditure and interest charges but before taxation.
- Taxation: Tax levied on taxable profits at a rate dependent on the relevant tax authorities.
- PAT: Profit After Tax.

12.3.6 *Cash Flow*

Understanding what happens to cash flow during an investment is essential. It is a method of appraising the investment and can have an effect on the rest of the business. For example, the business may need to borrow funds for a defined period, with the associated financing charges, or self-fund out of its cash reducing the cash balance and possibly reducing other income such as interest received. Table 12.5 shows the cash in and out for the first five years of the business case. Year 6 is the same as year 5. The cash balance is derived from the PAT, adding back non-cash items (depreciation). Here, capital expenditure (CAPEX) is all incurred in the first year and reduces the cash balance at this point.

Figure 12.2 shows the annual and cumulative cash flow for the business case. The simple cumulative cash flow is generated by adding the annual future cash flows together and shows the investment paying back in year 5. However, this does not reflect the time value of money ($100 in five years is likely to purchase less than $100 today), nor the risk that the future cash will not materialise. By applying a discounted cash flow (DCF) analysis to these future cash

Table 12.5 Cash flow ($) for a model factory/foundry

	Year 1	Year 2	Year 3	Year 4	Year 5
Cash in					
Revenue		62,238,155	112,214,722	136,706,697	136,706,697
Cash out					
CAPEX[1]	127,145,987				
Total costs	21,332,516	41,875,635	56,623,578	63,997,549	63,997,549
Taxation[2]				16,386,344	21,812,744
Net cash/annum	−148,478,503	20,362,520	55,591,144	56,322,804	50,896,404
Cumulative	−148,478,503	−128,116,001	−72,524,857	−16,202,053	34,694,351

[1] All CAPEX is paid for at the end of year 1.
[2] Tax is paid during the year after it was incurred, that is, the tax paid in year 5 was incurred in year 4 (see Table 12.4). It is also assumed that the loss made in year 1 (see the Revenue–Costs line in Table 12.3) can be carried forward to offset future profits. Hence, the first tax relates to year 3 and is paid in year 4.
Abbreviation: CAPEX, capital expenditure.

flows, it is possible to address both issues (Whole Life Forum, 2012). The discount rate used in the business case is 10%.

The net present value (NPV) gives the present value of a future cash flow, based on an assumed discount rate. For example, the NPV of the Company X business case in year 5 is −$6.3 million and in year 6 is $22.4 million, as can be seen in Fig. 12.2, which shows these flows graphically. A related investment metric is the internal rate of return (IRR), which is the rate at which the NPV is zero. This is effectively the percentage return on investment (ROI) for this business case. Company X can then compare this IRR with returns possible with other investment options.

Table 12.6 shows an extract from the undiscounted cash flow, NPV and IRR of the business case in years 4, 5 and 6. If the calculated return were acceptable, then the investment would go ahead.

The formula for the NPV is

$$\mathrm{NPV} = \sum_{i}^{n} \mathrm{CF}/(1+r)^{i} \tag{12.1}$$

where:

CF_n = cash flow in year n
r = discount rate

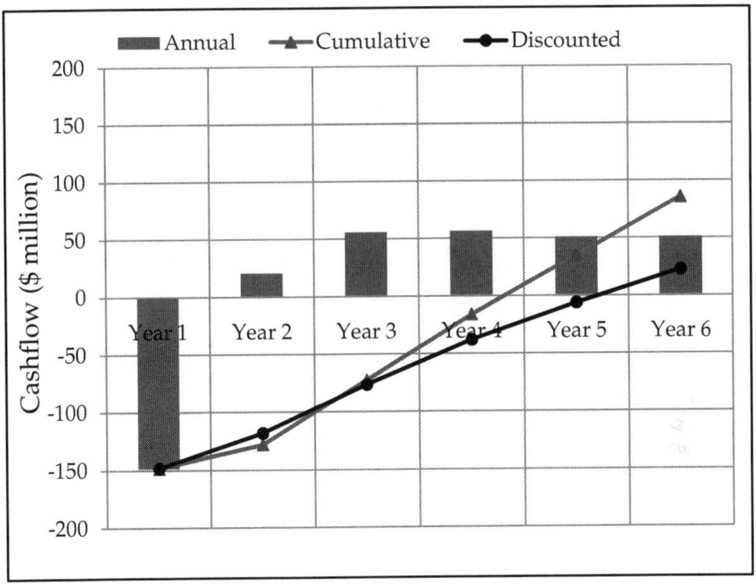

Figure 12.2 Cash flow for a model factory/foundry (Table 12.5).

Table 12.6 Extract of cash flow metrics

	Year 4	Year 5	Year 6
Undiscounted cumulative Cash flow	−$16.2 M	$34.7 M	$85.6 M
NPV	−$37.9 M	−$6.3 M	$22.4 M
IRR	−4.9%	8.0%	15.7%

12.4 Depreciation

Depreciation is a means of allocating the cost of a fixed asset over its useful economic life. It matches the estimated cost of using the asset with the benefits received. Depreciation also reflects the decline in the asset's value in the balance sheet over its useful life (tutor2u, 2012).

It is important to note that depreciation itself is a non-cash item—the cash element is the original purchase of the asset—and that the 'useful life' is an estimate and assets may remain in use subsequently. Fixed assets and their subsequent depreciation are reflected in the financial statements.

There are several methods of calculating depreciation, but the simplest is the linear depreciation method, where the purchase value of the asset is simply divided by its estimated useful life (see Table 12.1).

If an asset is expected to have a scrap value at the end of its nominal useful life the depreciation is calculated from the capital cost of the asset, less its residual value divided by its nominal life.

$$r = (C - S)/ \tag{12.2}$$

where:

r = annual depreciation
C = capital cost
S = scrap value
Y = number of years

Table 12.7 shows the depreciation cost, the annual book value of the asset and the annual depreciation for a $100,000 asset which, after five years, ends up with a scrap value of $10,000.

MEMSCOST classifies fixed assets into buildings, plant, clean rooms and equipment. Each of these can be individually selected and populated in MEMSCOST. This means that depreciation for each type of fixed asset can be included or excluded from the calculation and each type of fixed asset can have its own depreciation period in years. Typically, buildings are depreciated over 25 years, a plant over

Table 12.7 Linear depreciation for a $100,000 asset, with a scrap value of $10,000 after five years

Year	Value	Annual depreciation	Book value
0			100,000
1	100,000	18,000	82,000
2	82,000	18,000	64,000
3	64,000	18,000	46,000
4	46,000	18,000	28,000
5	28,000	18,000	10,000
Total	**10,000**	90,000	

15 years, clean rooms over 10 years and equipment (e.g., fabrication machines) over 5 years.

12.5 Profit, Profit Margin and Markup

Revenue (R) is the price per device multiplied by the number of devices.

Profit (P_R) is the difference between the revenue (R) and the total costs (C), and the ratio can be expressed either as a 'profit margin' or as a 'markup' (entrepreneur, 2000).

The profit margin (P_{MAR}) is the profit divided by the price and is the most common form quoted, that is,

$$P_{MAR} = (R - C)/R \tag{12.3}$$

The markup (P_M) is the profit divided by the cost, that is,

$$P_n = (R - C)/C \tag{12.4}$$

By combining Eqs. 12.3 and 12.4, the relationship between profit margin and markup can be seen (Fig. 12.3):

$$P_{MAR} = P_n C/R \tag{12.5}$$

12.6 Total Cost of Ownership

The total cost of ownership (TCO) is an analysis tool used to calculate the overall cost of owning an asset over its defined useful life (Munzio, 2000; Sematech, 2002; Sematech, 2007). Costs include, but are not limited to, the costs of purchasing, operating, maintaining and developing the asset, which may be direct or indirect. Although the TCO enables the true economic cost of an asset to be estimated throughout its lifetime, it has limitations. By focusing purely on the cost of owning an asset, it does not factor in the benefits of the asset, which may drive the purchase decision towards a cheaper but potentially less effective option. More expensive options that may be necessary to meet more demanding applications or give a better ROI can be overlooked.

Table 12.8 TCO for a wafer aligner (all costs in $)

Item	Capital	Depreciation 5 years	Throughput wafers/hour	Utilisation 86%
Capital equipment	250,000	50,000	40	0.165
Consumables	112,500			4.500
Masks (1.0 μm)	3,066			0.123
Maintenance	11,178			0.447
Labour	28,738			1.150
Utilities				
Plant (capital)	62,500			4.615
Electricity	52,873			
Space				
Clean rooms	67,200			2.946
Plant	6,444			
			Cost/wafer	13.94
			Cost/hour	0.349

The TCO should be calculated over a defined period, for example, a year or the lifetime of the asset.

$$\text{TCO} = \text{Capital} + \text{Maintenance} + \text{Consumables} + \text{Plant} + \text{Floor space} \quad (12.6)$$

The maintenance costs include the replacement parts + consumables + manpower required to keep the equipment operational. A simple maintenance model is assumed, where the parts and consumables are fixed percentages of the capital cost and the manpower is determined by routine and breakdown maintenance. In practice these figures will vary from installation to installation, from machine to machine and from one maintenance regime to another.

The estimated TCO for a wafer aligner is shown in Table 12.8, including the items defined in Eq. 12.6. The TCO is considered on a per wafer or per hour basis.

12.7 Return on Investment

The ROI must be carefully defined (entrepreneur, 2012), particularly in the period over which it is calculated.

The ROI is calculated from:

$$RoI = P/C \qquad (12.7)$$

where:

C = cost of the investment
P = net profit from the investment

For example, if a $100,000 investment in a new machine generates a $50,000 net profit from the investment in a short period, for example, 1 year, then the ROI = 50% over 12 months. The main problem with the concept of ROI is deducing appropriate income and returns over a given period.

12.8 Earnings per Share

An indication of a company's profitability is the earnings per share (EPS) (Investopedia, 2012), that is,

$$EPS = (I - D)/S_{\mathrm{H}} \qquad (12.8)$$

where:

I = net income
D = dividends paid to shareholders
S_{H} = average of outstanding shares

Consider the company's PAT of $32.87 million shown in Table 12.4. The company decides to allocate $3 million to the shareholders, of which there are, on average, 10 million. Then the EPS = ($32.87 – 3)/10 = $2.99 per share.

References

entrepreneur (2000). Gross profit and markup.www.entrepreneur.com/article/21936.

entrepreneur (2012). Return on investment.www.entrepreneur.com/encyclopedia/term/8270.html

IFRS (2012). International Financial Reporting Standards 2012 'Red Book'. www.ifrs.org/Features/Redbook.htm

Investopedia (2011). Understanding the income statement. www.investopedia.com/articles/04/022504.asp axzz 2440 E5YM.

Investopedia (2012). Earnings per share – EPS.www.investopedia.com/e/eps

Munzio, E (2000). Optical lithography cost of ownership (COO) calculations – final report for LITG501. *International Sematech Technology Transfer 00104614A-TR.*

Sematech (2002). Fab utility cost values for cost of ownership (COO) calculations. *International Sematech Technology Transfer 0203426A-TR.*

Sematech (2007). A cost of ownership (COO) model for Probe cards version 7.1. *International Sematech Manufacturing Initiative.*

tutor2u (2012). Business financing and accounting. Depreciation of fixed assets. tutor2u.net/business/accounts/assets_fixed assets_depreciation.asp

Whole Life Forum (2012). Net present value and discounted cash flow. www.wlcf.org/npv.html.

Chapter 13

Conclusions

The objective of this book is to take the reader from processes and equipment that are commonplace in a microelectromechanical systems (MEMS) research and development (R&D) laboratory, where the demonstration of principles is paramount, to their application in an industrial environment, where cost may determine the success of the product in the marketplace.

This book describes an approach and methodology for the costing of MEMS. The MEMSCOST Excel spreadsheet designed by the author (see Chapter 8) is used as the vehicle for the cost calculations contained herein. Figure 13.1 summarises the basic methodology adopted by MEMSCOST.

As with any calculations, the output of MEMSCOST is sensitive to the quality of the input information and the costing in this book is based on best estimates. Revenue and price data is taken from publically available market surveys, and the cost data comes from both publically available data and the author's experience and interest in the costing of MEMS technology over the last 35 years. Where more accurate information is available, for example, for a particular factory, MEMSCOST inputs may be customised. The combination of MEMSCOST costing and the available pricing data points to profit margins of approximately 40%.

MEMS Cost Analysis: From Laboratory to Industry
Ron Lawes
Copyright © 2014 Pan Stanford Publishing Pte. Ltd.
ISBN 978-981-4411-06-6 (Hardcover), 978-981-4411-07-3 (eBook)
www.panstanford.com

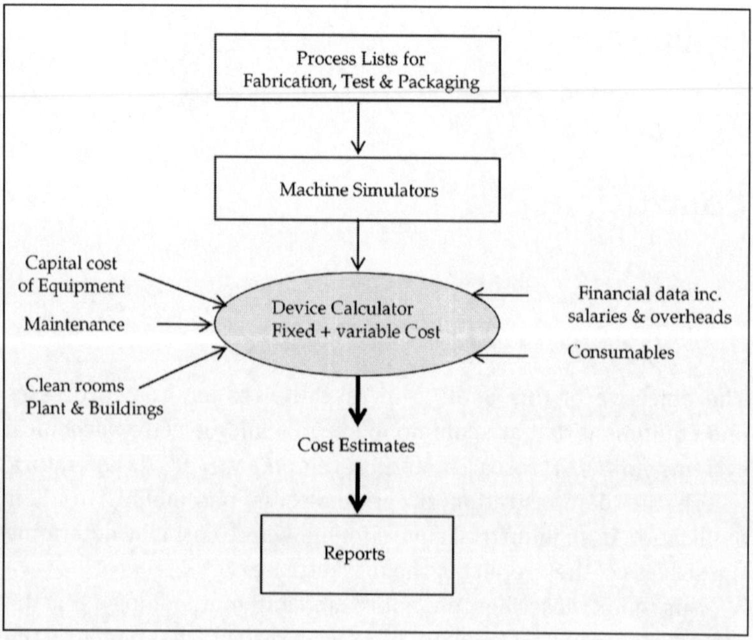

Figure 13.1 The basic structure of a cost analysis spreadsheet (MEMSCOST).

The manufacturing processes and techniques for MEMS are described in sufficient detail for costing to be undertaken, with fabrication in Chapter 2, testing in Chapter 5 and packaging in Chapter 6. The equipment and processes have been described for surface micromachining thin films in Chapter 3 and high-aspect-ratio microstructures in Chapter 4.

The engineering details required are in a process list that itemises the process steps to fabricate, test and package the product and an estimate of the throughput and capital cost of each machine associated with a process step. Throughput can be calculated using a machine simulator, examples of which are given for each machine type. Maintenance levels can be set and an estimate made for the clean rooms, buildings and plant (Chapter 7). Financial and management data is automatically included, for example, salaries, depreciation and overheads, unless specifically excluded.

The methodology requires fixed and variable costs as input and provides quantitative output, with costs dependent on wafer and die size as well as the number of units manufactured and sold per year.

The devices used to demonstrate the capabilities of MEMSCOST are accelerometers (Chapter 9) and microphones (Chapter 10), which are based on silicon as a substrate, as this material is expected to dominate the large-scale production of MEMS by integrated device manufacturers (IDMs). However, there are important MEMS applications, such as microfluidics, where non-silicon materials (polymers) are used, where laser ablation is the fabrication technology of choice and where injection moulding may be a key technology for replication. The same costing methodology will give good estimates, provided the equipment and process parameters are well documented.

A significant proportion of MEMS fabrication has moved towards standard integrated circuit (IC) equipment and processes. However, this is less the case with testing and packaging due to the conflicting needs of MEMS. For example, MEMS accelerometers require impact testing and packaging that seals the MEMS component from the environment (a similar requirement to IC packaging), while MEMS microphones require acoustic stimulation and packaging that provides access to the environment.

As with IC manufacture, MEMS ideally requires wafer-level and post-packaging testing for both parametric and functional testing. Testing of MEMS has made some progress in recent years, and some equipment has become available from specialist automatic test equipment (ATE) manufacturers, but much of the detail still remains company-confidential. In the absence of comprehensive capital and throughput data, MEMSCOST uses the expectations contained in the International Technology Roadmap for Semiconductors (ITRS, 2011). Over the next few years it is anticipated that more data will become available.

For some years, packaging was estimated to be by far the most expensive component of the total cost of manufacture (e.g., 80%). However, with the advent of standardisation with IC-like packages for high-volume MEMS devices and more accurate cost analysis, packaging cost estimates have been much reduced. For example, testing and packaging costs may now be less than two-thirds of the total cost of manufacture of a MEMS device. Nevertheless, the manufacture of some MEMS devices (e.g., microfluidics) will continue to need non-standard, relatively expensive packaging solutions.

Foundries play an important role in supplying the industry with essential MEMS manufacturing capability. They offer expertise in specialist processes, such as deep reactive-ion etching (DRIE) and excimer laser ablation for microfluidics and provide R&D and small- to medium-scale production. More recent developments have seen large-scale, pure-play IC foundries enter the MEMS field (e.g., Taiwan Semiconductor Manufacturing Company [TSMC]).

MEMSCOST is designed to compute the cost structure of a device from an engineering perspective. However, the illustrative business case in Chapter 12 shows that MEMSCOST data can be used to construct simple financial statements and financial assessments of any proposed investment. It also looks at several key financial metrics, such as total cost of ownership.

MEMSCOST is not commercially available. However, cost analysis packages are available from two companies both offering MEMS and IC versions. IC Knowledge introduced its IC cost model in 2001 and its MEMS cost model in 2004. SystemPlus introduced the manufacturing tool MEMS CoSim+ in 2009.

It is inevitable that MEMS-based devices will experience further integration of functions, where both technology and cost will be paramount. For example the 10 degree-of-freedom devices will require three-axis accelerometers, three-axis gyroscopes, three-axis magnetometers (compass directions) and a pressure sensor for altitude. Currently these functions are available only as separate die, often with the signal-processing ASIC and the MEMS sensor also on a separate die, albeit in a single package. Integration is only possible at the board level.

The initial approach may be to develop multi-die assemblies where more than one function is contained within a package. Further integration by combining several functions at the die level, for example, accelerometers and gyros, could reduce both fabrication and packaging costs. This trend suggests that the predominant manufacturing technology will be based on CMOS-

MEMS and cost analysis would be an important tool to help determine the most cost-effective manufacturing technologies.

The last 12 chapters have taken the reader through all the key aspects of costing MEMS, from physical structures to the technology of manufacture, including fabrication, testing and packaging the completed device. Capital investment for equipment, plant, clean rooms and buildings has been shown to require major investment, from tens of millions of dollars for a MEMS factory to billions of dollars for an IC semiconductor facility.

In conclusion, it has been shown that a 'bottom up' costing process, based on detailed knowledge of the equipment and processes used and typified by the MEMSCOST approach, can produce results that are in broad agreement with top-down analysis typified by road maps and by market analysis. Cost analysis for MEMS is now a sufficiently robust process to be an integral part of design and manufacture. It bridges the gap between engineering and finance.

Index